To the memories of Robert H. Lawrence, Elliot See, Charles Bassett, Vladimir Komarov, and the crews of Apollo 1, Soyuz 11, *Challenger*, and *Columbia*. Your tales will never be forgotten.

CONTENTS

ACKNOWLEDGMENTS

I cannot possibly thank all of those who kindly gave their time and knowledge to this project. Just know that I am very grateful. I do need to send special thanks to Jake Bonar with Prometheus Books for his belief in this book and for going to bat for me to make it happen. Thanks also to acting NASA Chief Historian Brian C. Odum and NASA HQ History Program Senior Archivist Colin A. Fries. Your insights into many of the stories included in these pages were invaluable. Thank you to two of my best friends, Debbi O'Connor and Jay Karpowich for their input and enthusiasm when I started writing this book. You got me off on the right foot. And to the best proofreader in the world, my wife, Maggie, who waded through the pages during its roughest moments and put up with hours and hours of my chatter when I discovered new tales to tell and couldn't stop talking about them. Thank you!

INTRODUCTION

I was four years old in 1962, living in a quaint suburban home in northern New Jersey. I clearly remember my mother ironing clothes in the living room, our behemoth of a Magnavox console television tuned to WCBS out of New York City. A grainy black and white image flickering on the screen caught my eye and stopped me in my tracks. A voice came out of the set's humongous cloth-covered speaker. The voice was counting backward.

I remember sitting down cross-legged in front of the set, intrigued by what was happening. What was this? Why was the man counting? On the screen, a sleek silver cylinder stood majestically amid clouds of smoke—or what I thought was smoke—venting from its side. When the man's countdown reached zero, a small but bright flame shot down from the lower sides of the cylinder. Seconds later, the bottom of the cylinder burst into a bigger white flame (it was black and white, after all).

Slowly, the cylinder climbed into the sky. I watched until the machine became a simple white dot on the screen. My mother explained to me that it was a rocket sending a man into outer space. I didn't realize that it was a Mercury mission or who the astronaut was on board, but it was fascinating.

I had seen rockets carrying men—and women—into space before, but those were fictional "supermarionettes" like *Fireball XL-5* or the animated *Space Angel*. This was the real deal, and I was hooked.

As I grew older, I would leave late for school purposely—or even miss a day—to watch the launch of the two-man Gemini missions. Another local tristate area channel, WNEW in New York, showed a fifteen-minute NASA documentary it called *NASA Presents* every morning just after signing on, and I would be up early watching it, possibly the only viewer in north Jersey tuned in. On any given Sunday morning, two of my best friends and I would find ourselves in the middle of a huge field at a local college launching a multitude of Estes model rockets, one after the other.

Although I was fascinated by space and the US space program, I wasn't your standard rocketry nerd. Yes, the amazing feats our astronauts performed hundreds of miles above Earth were spectacular, but I was more interested in the stories *behind* the story. The little tales that flew under the radar of most mass media outlets and often showed up in section E at the very back of local newspapers: a missile tested by the German rocket team led by Dr. Wernher von Braun in New Mexico went awry and blew up a cemetery in Mexico, giving the team the notoriety of being "the only German team to attack Mexico from its bases in the United States"; the day the "cola wars" made it into outer space; the man who bought a plot of land on the moon's Sea of Tranquility and later sued NASA for moon rocks that he believed had been removed from his property.

That's where the idea for *Space Oddities* was born. I wanted to bring to life some of those little-known or forgotten stories—pranks pulled by NASA astronauts and engineers, superstitions that mission controllers bring to their consoles to "guarantee" a successful mission, product marketing in space, including the "space pens" from the Mir space station promoted on the cable shopping channel QVC. The stories go on and on. Sometimes they are funny and make you chuckle. Sometimes they are sentimental and bring warmth to your heart. Other times, they are tragic and bring a tear to your eye.

Now some of you may recognize a story or two within these pages. I think you will find that even those tales have some fascinating and surprising twists and turns—I was surprised by some of the stories I thought I knew.

Writing *Space Oddities* was a true labor of love and I hope my offerings keep you enthralled as I was doing the research and writing for this book. Thank you for joining me on this ride. It's time to tell a story.

1

THE ROCKET WORKED PERFECTLY EXCEPT FOR LANDING ON THE WRONG PLANET

As you would expect, the early days of rocketry and spaceflight were fraught with setbacks, disappointments, and fatalities. Although there are many accounts of rocketlike devices being built throughout ancient history (mostly by accident), it is safe to say that the Chinese were the first to put gunpowder to use as an early form of rocket propellant. The first documented rocket is believed to have been the Chinese fire arrow,[1] which was basically a bamboo tube loaded with a mixture of saltpeter, charcoal, and sulfur that was then attached to an arrow. The archer placed the arrow on his bow, lit the mixture, and let it fly in the hopes of vanquishing his rivals during battle.

The first recorded use of the fire arrow was found in the document *Wu-Ching Tsung-Yao*[2] by Tseng Kung-Liang in 1050 AD. It wasn't long after the Kung-Liang document appeared that warriors realized that the bow was not needed to launch their arrows. They could simply ignite the arrow and it would fly on its own. Today, we would call these bottle rockets, albeit on a much smaller scale. No matter what you call it, the first rocket was born.

The innovation of fire arrows led to a legend that has withstood the test of time and one that labeled a local Chinese government official and stargazer, Wan Hu, as the world's first astronaut.

The legend of Wan Hu[3] dates to around 1500 AD during the Ming dynasty. Wan Hu was obsessed with the stars and dreamed of one day traveling to them. He envisioned using the fire arrow to propel himself into the

Wan Hu attempts to become the first "astronaut" around 1500 AD. *Illustration courtesy of United States Civil Air Patrol, public domain, via Wikimedia Commons*

heavens but realized that a single fire arrow alone would not be able to lift a heavy man. He deduced that if he strung several fire arrows together, they just might create enough force to lift him off the ground.

Wan Hu's half-baked scheme involved hauling out his ornate bamboo throne, affixing two kites to it (apparently for aerodynamic stability), and encircling the chair with forty-seven of the largest fire arrows he could find. Dressed in his finest imperial robes, Wan Hu sat down in the chair and ordered forty-seven of his servants, each armed with a flaming torch, to light the arrows.

The servants stepped forward and ignited the arrows. An enormous and deafening explosion rocked the area, and a mighty cloud of smoke enveloped the "launch" site. When the smoke had cleared, Wan Hu and the chair were gone. Wan Hu had disappeared, leaving onlookers to believe that he actually had been hurled into space, which would have made him the world's first astronaut.

That's not quite how the story really ended. In an episode of the Discovery Channel's show *Mythbusters*, hosts Adam Savage and Jamie Hyneman re-created the Wan Hu "spacecraft" with their crash-test dummy Buster playing the role of Wan Hu. The nasty explosion resulted in extensive damage to Buster and what would be fatal burns for a real human. Even though the legend could be disproven easily, a crater on the far side of the moon was named in honor of this rocket daredevil in 1970.

Legends aside, early visionaries, scientists, and dreamers recognized that flight to the stars would require some sort of rocket power. Prior to 1930, there had been many experiments in rocketry using various combinations of chemicals and gases as fuel, which resulted in varying degrees of success. These early rockets were all unmanned. No one had ever attempted to ride one. Enter Max Valier (pronounced *Val-yay*).

Valier was born in Bozen, Austria-Hungary (now the town of Bolzano, Italy), on February 9, 1895. At the age of thirteen, Valier's father died, and the young man and his mother moved in with his grandparents. It was here that the inquisitive young man discovered two old telescopes that his grandfather had received as collateral for an unpaid loan he had granted. Valier took it upon himself to rebuild the telescopes, and his love of astronomy began.[4]

Valier attended college at the University of Innsbruck, where he majored in math, physics, and astronomy and minored in meteorology. It quickly became obvious that Valier was more of a showman and promoter than scientist. His head was literally in the heavens as he dreamed of traveling to the stars, and he began a journey that later earned him the moniker of rocketry's chief cheerleader.

The public experiments and presentations that Valier conducted were literal flights of fancy. In 1914, for instance, the townspeople of Innsbruck were frightened when they spotted a blazing comet overhead that appeared to be heading straight for the city. It turned out that the comet was actually one of Max Valier's "experiments." A report in the *Tiroler Nachtrichten* newspaper documented the incident.

> [Valier] constructed a kite out of impregnated paper, a kind of glider, to which he had attached a system of three fireworks instead of a tail, in such a form that one rocket lit the second and then the third before it burned to the end. Start: Hotel Mariabrunn on the Himgerburg. The kite landed in Pradl. . . . No one could argue quite seriously that Max Valier is [not], in fact, the inventor of the jet aeroplane.

Valier's family convinced local police not to arrest the young man and to keep the incident quiet so he could continue his studies.

The advent of World War I saw Valier join the military, and his meteorology degree landed him at a weather station on the front. During this time, Valier survived the crash of both a weather balloon and a plane. He also earned the distinguished Austrian Silver Star after taking out a machine that was used by the enemy to spew mustard gas, which saved the lives of countless soldiers.

Following the war, Valier married a divorcée, Hedwig Alden, and returned to college, studying in Vienna and Munich, but he never completed his advanced degree. Still, he was enthralled with the possibilities of rocketry and became more interested in writing and popularizing those possibilities by painting incredible images through his words of futuristic innovations than in completing his studies.

In 1923 in Germany, Valier picked up a copy of a book that would change his life: *The Rocket into Planetary Space* by German rocket pioneer Hermann Oberth. Valier was so absorbed by the book that he wrote to Oberth, and with the rocket pioneer's help, he penned his own booklets: *Der Vorstoss in den Weltraum* (*The Advance into Space*) and *Raketenfahrt: Eine Technische Moglichkeit* (*Rocket Travel: A Technical Possibility*). Both publications expanded Oberth's vision of what space travel could be and became wildly popular, helping to spread the excitement of rocketry and its possibilities throughout the country. Its success prompted Valier to continue writing. His magazine and newspaper articles were published more and more frequently and their popularity increased exponentially, fueling rocket fever across Germany with their vivid pictures of futuristic rockets and space planes.

During this time, Valier began experimenting with solid fuel rockets and proposed developing a rocket-powered car. In 1927, with funding by automobile pioneer Fritz von Opel and a solid propellant engine developed by Frederick Sanders, Valier took a standard Opel automobile and fitted it with two solid rockets. It was called the RAK 1 (a shortened form of *rakete*, the German word for rocket). With the massive worldwide public relations department at Opel behind him, word of the experimental car spread like wildfire across the globe.

On April 11, 1928, with Opel test driver Kurt C. Volkhart behind the wheel, RAK 1 was ignited. The test vehicle belched a thick cloud of smoke and took off down Opel's test track, accelerating from zero to sixty-two miles per hour in only eight seconds. The Scottish newspaper, the *Dundee Courier*, wrote:

> [The car] was not driven by a petrol engine but by a so-called rocket about which details cannot be disclosed, but it can be said it was constructed in accordance with the plan of the German Max Valier, known as the "Fantastic Cosmos Flier" because of his idea of building a rocket which could be sent out into space. We are convinced that the "Opel Sander Aggregate," as the new machine is called, will achieve a speed hitherto considered impossible on the surface of the earth, and it will prove to be only a step toward the construction of a rocket air machine and to a cosmos airship on the lines of the Valier project.[5]

That same year, Valier organized the Verein fur Raumshiffahrt (VfR), or space association, the world's largest rocketry society, with an impressive membership that included Oberth, science writer Wily Ley, and a young Wernher von Braun. Valier also continued building on his "rocketry cheerleader" label by racking up more and more impressive successes: In 1928, his rocket car reached 145 miles per hour, and in 1929 he built the first rocket sled, which would reach an incredible speed of 250 miles per hour.

Valier began experimenting with liquid fuel rockets that used a mixture of kerosene and liquid oxygen. The highly volatile combination would be pressurized and funneled into a steel cylinder where it would ignite. The flame's exhaust would shoot out of a nozzle in the cylinder, which was mounted in a rudimentary test stand.

The experimenter had no regard for his own safety. He was focused only on results and never once took safety precautions such as standing behind protective blast-proof barriers or in bunkers. He never wore goggles or gloves and sat only a few feet away from the engine's exhaust to manually control the pressure in the cylinder.

Valier and his assistant Walter Riedel fired up his first such engine on May 17, 1930. The results were impressive, producing more thrust than they had anticipated. After taking a short break, the euphoric pair fired up the engine a second time, and once again, the total thrust produced was much more than they had dreamed possible.

The exuberant Valier wanted to light up the engine one more time. He pressurized the chamber and ignited the engine. A mighty explosion rocked the room. According to accounts in the *New York Times*, a ten-foot flame enveloped Valier, throwing him twenty feet across the room. When Riedel reached Valier, he found the inventor's body mutilated by shrapnel. He died within ten minutes, becoming the first recorded death in modern rocketry.

Despite his death, the excitement that Valier had generated across Germany about the incredible promise rockets held for mankind only intensified and one of his fellow VfR members, a young Wernher von Braun, picked up the mantle to continue pushing the envelope of rocketry even further while at the same time bringing the public along for the ride through a concentrated public relations campaign. His natural abilities in this arena would serve him well throughout his career.

Much like Valier, von Braun read the Oberth book, which ignited his passion for spaceflight. Unlike Valier, however, the book prompted von Braun to earn an advanced degree in calculus and trigonometry so that he could learn more about the intricacies of rocket mechanics.

In 1932, Germany was still abiding by the Versailles Treaty of Paris, which limited the type of weaponry the country could develop to defend itself from foreign attack. German army artillery captain Walter Dornberger paid a visit to the VfR to see if rockets could be used in this capacity without violating the treaty.

The group of young scientists held a demonstration for military brass and immediately underwhelmed them when the test rocket failed miserably. Dornberger, however, was so impressed with the intelligence and abilities of von Braun that he asked him to head the army's new rocket unit while

Historic view of Max Valier in an early engine static test firing. The rocket is sitting on a scale. Valier is measuring thrust by adding weight like the one in his right hand. *Courtesy of the Library of Congress, HAER ALA,45-HUVI.V,7A—19*

continuing to work toward his doctorate in physics. It didn't take von Braun long to decide—he accepted the position, seeing the military as a means of furthering his quest for developing larger and more powerful rockets that would one day send a man to the moon.

Two years later, von Braun's team of eighty engineers developed its first prototype liquid fueled rocket called the A2 (aggregate rocket 2), which showed a lot of potential during two successive public demonstrations. It was time to move on to a larger vehicle.

Work immediately began on the A3 and A4 rockets. Whereas the A3 gave the team valuable insight into guidance control, its sibling, the A4, was stubborn, suffering a remarkable series of failures during the next several years. When World War II began, it is estimated that almost three thousand A4 rockets were tested, all built with slave labor from Nazi concentration camps.

By 1944, the war in Europe was nearing its end. In one year, the Third Reich would be destroyed, but not before Hitler gave the order to turn the A4—which was renamed V-2, or vengeance weapon 2—loose on England. On September 8, 1944, the first V-2 flew more than two hundred miles, lifting off from the Hague-Wassenaar region of Holland and reaching the edge of space before crashing down with horrible devastation in London, killing three and injuring seventeen. Von Braun's dream of using his rocket to send humans into space had been turned into a killing machine by the Nazis and he was heard to lament, "the rocket worked perfectly except for landing on the wrong planet."[6]

One year later, the writing was on the wall. Germany was in dire straits as the war ended. The allies were moving in—the Russians from the east and British and American forces from the south and west. Von Braun had to decide whether to be captured by the Russians, facing torture from Stalin's regime or perhaps even being shot on sight, or to surrender to the Americans. He chose the latter.

Von Braun's brother Magnus rode his bicycle across the Austrian border where he encountered the US Forty-Fourth Infantry Division. Magnus explained that a large contingent of German rocket scientists including his brother Wernher were ready to surrender to the Americans. The commander didn't believe his story and told him to return with proof.

Wernher and several of his colleagues returned to meet with the commander and were immediately escorted to the Counter Intelligence Corps

(CIC) headquarters in Reutte, Austria, where more than five hundred of the scientists were interrogated to determine which of them would be allowed entrance into the United States.

Von Braun was interrogated for three days by Lieutenant Walter Jessel. On the final day of questioning, Jessel looked at the German and asked, "Dr. von Braun, if you were to emigrate to the United States and become a naturalized citizen, what would you do with the rest of your life?"

The young rocket man didn't hesitate and replied, "I would build a rocket and fly to the moon."[7]

In what was originally dubbed Operation Overcast, most of the rocket team, along with the remaining V-2 rockets and components, were officially surrendered to the Americans. As he did when he accepted Dornberger's invitation to join the German military effort to design rockets, von Braun called the surrender a fair trade-off: helping the Americans with their faltering military missile program while at the same time keeping his longtime dream of sending men into space alive.

It's interesting to note that following close behind the Allied forces as they marched into Germany was a contingent of soldiers known as the Combined Intelligence Objectives Sub-committee (CIOS), which combed the area for war-related documents and materials. One discovery was found in a toilet at Bonn University. It was a catalog that listed the names of Third Reich scientists and engineers. With this information, CIOS was able to round up an additional sixteen hundred German scientists who joined their colleagues in America to work on various projects for the US military.

However, getting von Braun and his team into America was a delicate public relations operation. After all, these scientists had worked for the Nazis to create a devastating weapon, the V2, which killed scores of people and caused millions of dollars in damages. The US Army's public relations team went into operation, telling the public that the team was a group of German rocket scientists who had "volunteered" to come to the states and work for a "very modest salary."

Despite President Truman's ban on former Nazis entering the country—let alone working for the government—the Office of Strategic Services (OSS), the forerunner of the CIA, whitewashed the scientists' records and sent the teams to bases in White Sands, New Mexico, and Fort Bliss, Texas, where they would continue to evaluate and improve on the V-2 technology. Test firings of the rockets would eventually include scientific experiments to study the upper atmosphere as well as the effects of spaceflight on small biological subjects including monkeys. The program was renamed Operation Paperclip, which was derived from the fact that the dossiers on the rocket scientists were paperclipped together.

In all, sixty-seven V-2 rockets were launched from the team's base in the New Mexico desert. Most of the launches were successful, not only putting the United States on track for developing long-range ballistic missiles but also taking mankind's dreams of flying off into outer space one step closer to becoming a reality. One of those rockets, however, nearly caused an international incident.

To the American public, the US government presented von Braun's team not as prisoners of war but as prisoners of peace. The media began referring to them as the "Germans who invaded America."

Test launches of the V-2 continued at an incredible pace; some, as we will later see, included early flights with live passengers ranging from mice to rhesus monkeys. On the morning of May 29, 1947, the team from Operation Paperclip prepared for yet another test flight. As the countdown reached zero and the launch command was given, an orange flame burst from the rocket's engine and it roared off the pad. Immediately, the missile began its preprogrammed arc as it rose into the cloudless cerulean desert sky, only this time, the rocket's guidance control system failed, causing the rocket to veer dramatically off course toward the south. Engineer Krafft Ehricke described what happened next:

> Fortunately, it was fast enough to arc over El Paso—I shudder to think what would have happened if it had hit El Paso—and it hit [Juarez, Mexico], and if memory serves, it impacted into a cemetery. We had already been called the only German task force that managed to invade the United States territory and penetrate the United States as far west as El Paso. Now, we were known as the only German team to attack Mexico from their bases in the United States.[8]

In 1950, von Braun and his team were transferred to the small, sleepy north Alabama town of Huntsville to assist with the army's guided missile development program, the Army Ballistic Missile Agency (ABMA). Once again, von Braun's team was sidetracked in its quest for spaceflight as it was tasked with the development of intermediate-range ballistic missiles (IRBM) for the military. The team took the knowledge gained from the V-2 and applied it to a new missile, the Redstone, and even though the rocket was designed for weapon deployment, the team submitted a proposal to Washington to use the Redstone to send the first satellite into orbit. When the offer was given the cold shoulder, the team continued with its work undaunted by rejection and improved on the Redstone design, making it even more reliable and able to deliver a nuclear bomb more than three thousand miles away. More importantly to the team, this new rocket, the Jupiter, could also hurl a small satellite into earth orbit.

At the same time, the navy was also developing a missile—the Thor. It was a rocket but only on paper, whereas the Jupiter had been built and would have a successful test launch on September 20, 1956.

In 1955, a commission was convened in Washington by Homer J. Stewart of the Jet Propulsion Laboratory. Stewart directed representatives of the army, navy, and air force as well commercial entities to discuss and determine who had the best plan to build a rocket that could send mankind's first satellite into orbit. On August 5, 1955, the committee voted five to two to give the honor to the navy and its plan to build a three-stage rocket called Vanguard. Once again, von Braun's team was left out in the cold, but they persevered and continued working quietly on the Jupiter.

In 1956, to the surprise of everyone, the Secretary of Defense, Charlie Wilson, issued a memorandum that gave the navy even more control over missile development. What became known as Wilson's Memo gave the navy the go-ahead to develop IRBMs while the army would develop short-range missiles, thus squelching further development on the Jupiter. Von Braun was even told by someone whom history describes only as a "distinguished gentleman" of Washington, "There is much talk about satellites these days. Satellites are not for you! I want you to keep your hands off!"[9]

The team was angered and frustrated over the government's attitude toward its project, especially since it already had a proven rocket ready to fly, but no one was as angry as Army Colonel John C. Nickerson Jr.

Nickerson was a career army man whose career included serving as the commander of a field artillery battalion in France during World War II, for which he was awarded two Silver Stars. Nickerson later recalled that while serving in Europe, he was fascinated by the sight of flares shooting off into the sky. Turns out, those flares were V-2 rockets. With his schoolteacher appearance and thick Southern drawl, Nickerson eventually worked his way up to becoming the project manager for the Jupiter program.

The night before Thanksgiving 1956, Nickerson headed to Capitol Hill to plead the case for the army's project, to no avail. Senior officials would not change their minds. With even more resolve, Nickerson returned to his office and penned an anonymous twelve-page response to Wilson, which he distributed to the Department of Defense, members of Congress, and the media.

To prove his point that the army was the correct choice for the project, Nickerson included in his document classified information that revealed tightly held secrets about launch dates and the design and performance of both the navy's Thor missile and the army's Jupiter missile. Nickerson also made the argument that Wilson's decision was based on corporate corruption, stating that Wilson had demonstrated nepotism by granting the project solely on the fact that parts for the rocket were manufactured by the company Wilson had once headed, General Motors.

In December, journalist Drew Pearson, who wrote for the syndicated column Washington Merry-Go-Round, received a plain manilla envelope in the mail with the words, "this may be of some use" handwritten across it.[10]

Realizing that he held classified information in his hands, Pearson immediately contacted the Defense Department, which seized it and began an investigation into who had leaked the documents. The investigation quickly led to John Nickerson.

On January 2, 1957, the army's inspector general, Lieutenant General David Ogden, brought Nickerson into his office for questioning. During the session, Ogden told Nickerson, "Whoever wrote it might find themselves charged with espionage."

Feigning ignorance, Nickerson replied, "How could that possibly be true?" to which Ogden replied, "You will find out."

That afternoon, Nickerson rushed back to his office, gathered the classified materials he had assembled, and set them ablaze in his fireplace. As the fire crackled away, military police burst in and placed Nickerson under arrest.

Nickerson was charged with perjury and fifteen counts of security violations. What was more serious, however, was that he became the first person in US history to be charged under the Espionage Act. The officer faced a $10,000 fine and forty-six years in prison.

At Nickerson's court-martial, the defense stated that the colonel had shown "loyalty to the Army and [his] country," adding that what he had done was "for the good of the nation." Among the defense witnesses was Dr. Wernher von Braun.

Eventually, the army dropped the espionage charge in exchange for Nickerson's guilty plea to fifteen counts of breaching army security protocols. Instead of forty-six years in prison, Nickerson was demoted for a period of one year, received a formal military reprimand, and was fined $1,500. Sadly, on March 2, 1964, both Nickerson and his wife Caroline were killed in a head-on collision near White Sands, New Mexico.

What happened next took the world by surprise and gripped it with fear. On October 4, 1957, the Soviet Union launched the first satellite, Sputnik 1. Surely if the Russians could orbit a satellite, then they could lob a nuclear bomb anywhere in the world.

That night, von Braun was working late when he received the news. That same evening, the rocket scientist had an invitation to attend a reception being held in honor of the incoming secretary of defense, Neil Hosier McElroy, at Redstone Arsenal, home of the army's missile program.

With his well-known confidence, swagger, and promotional chops, von Braun took the news of Sputnik with him to the reception and walked up to McElroy and said, "Mr. McElroy, When you head back to Washington, you'll find that all hell has broken loose. We can put up a satellite in sixty days once you give us go-ahead."

The commandant of the ABMA, von Braun's boss, Major General John Medaris, confidently turned to von Braun and gave the German rocket scientist the go-ahead to begin making preparations.

The world waited for the US response. Surely the Americans could counter this international threat with its technological prowess. But despite the successes of von Braun's team, the government once again selected the navy and its Vanguard rocket to launch the country's first satellite.

Vanguard had made it from drawing board to production and had two successful test flights before it was called upon to answer the call of the nation and the world. On the morning of December 6, only two months after Sputnik, the slim, pencil-thin rocket was readied for launch. Its payload was a tiny six-inch, three-pound satellite.

With the world watching, the launch command was initiated. A brief flare of flame shot from the rocket's engine, followed by a thick cloud of smoke. The elegant rocket slowly rose four feet above the launchpad before the main engine began losing thrust. The rocket began sinking back onto the pad. An enormous hellish red-and-orange fireball engulfed the rocket, making it look like it was devouring itself. Just seconds before the rocket was completely destroyed, the nose cone rocked and tumbled away from the ensuing explosion. The battered sphere now resides in the Smithsonian's Air and Space Museum.

The world was stunned, and Americans were embarrassed. Headlines in newspapers rang out the news: "Vanguard Rocket Burns on Beach,"[11] "Kaputnik," "Flopnik." Fingers were pointed, with one senator from New Mexico, Clinton P. Anderson, telling reporters that the government knew Vanguard wasn't ready for the big show. Vanguard's prime contractor, the Martin Company, saw its stock plummet.

To add insult to injury, several members of the Soviet Union's delegation to the United Nations approached the US ambassador, Arkady A. Sobolev, and asked if he was interested in accepting aid from the Soviet Union's "technical assistance to backward nations" program. When asked by the press about the offer, Sobolev only replied, "no comment."[12]

Following the Vanguard incident, the Washington "distinguished gentleman" sent an urgent telegram to von Braun. It simply read, "Can you launch a satellite quickly?"

Indeed he could. On January 31, 1958, at 10:48 p.m., the main engine of the Jupiter-C ignited and the rocket with its payload, the satellite Explorer 1, soared into the black Florida sky from Cape Canaveral. One hundred and eight minutes later, Explorer was in Earth's orbit.

One of von Braun's dreams had finally been achieved—using rocketry for the advancement of science, which eventually would lead to putting men into space and onto the moon.

He had put America firmly in the race for space, but the scientist wasn't at Cape Canaveral to view the launch. He was so confident of the rocket's success that he had flown to Washington to take part in a press conference celebrating the nation's first satellite with the world.

Years later, von Braun reflected on not being at the momentous launch saying, "When one has worked for twenty-eight years on such a thing and now expects its culmination, it was a fairly bitter pill to swallow."

2

HEY SKY, TAKE OFF YOUR HAT, I'M ON MY WAY!

As the death toll in Europe rose with alarming speed in early 1917 and merchant ships lost to enemy U-boats increased to unfathomable numbers, President Woodrow Wilson stood before Congress and asked for a declaration of "war to end all wars." With that, the United States entered World War I.

As thousands of men shipped out and headed off to the trenches of Europe, women were called to serve. From cities and towns across the country, more than twenty thousand women were activated and sent to the nation's heartland to keep food production moving forward while the men who had once worked America's farms were away. Known as the Women's Land Army of America (WLAA),[1] many of its volunteers had never seen a plow or tractor let alone operated one. The women were known as "farmerettes" and were paid the same wages as their male counterparts.

Although the war ended in 1918 with the men returning home to their farms, many within the WLAA wanted to continue the work they had started, but financial restrictions forced the program to shut down in 1920.[2] The women returned to their homes and were once again relegated to being housewives and homemakers.

Women were called upon in a similar manner during World War II when 350,000 of them volunteered for uniform duty in the Women's Army Auxiliary Corps (WAACS) as well as the Navy, Marine, and Coast Guard Women's Reserve and the Army and Navy Nurse Corps. Another 4.5 million became known as "Rosie the riveters" when they joined the defense industry working in factories building aircraft.

Prior to entering the war, the United States produced only 921 planes annually. With the help of the Rosies, that number swelled to more than 92,000 units annually by 1944, with a total of more than 300,000 planes being built by war's end. In an interview with the Library of Congress, Boeing line worker Inez Sauer said that her mother cautioned her about factory work. "My mother warned me when I took the job that I would never be the same," she said. "You will never want to go back to being a housewife. After the war, I couldn't go back."[3]

That was the attitude of many of the women who served their country during the war; there was no going back to their previous life and status. Women had proven to the world that they were up for a challenge and could do the job of their male counterparts, but as was the case following the previous war, most of the women returned to being housewives. When they did continue in the workforce, they found themselves playing second fiddle to their male counterparts, earning meager wages at best, and often relegated to minor positions.

The same attitudes toward women persisted as America ramped up its quest to send a human into space. Sending a woman into the cosmos was considered almost blasphemous to many of the men involved in the program, but there were those who looked beyond the stereotypes and genuinely believed that women could do the same job as men in space, if not better. One of those men was Dr. Randolph Lovelace.

Influenced by his uncle who had established a medical research facility in Albuquerque, New Mexico, Lovelace enrolled in college at Washington University in St. Louis at the age of nineteen, during which time he developed a passion for flying. He joined the US Naval Reserve Training Corps, eventually earning his wings at the Great Lakes Naval Station in Illinois.

Lovelace continued his education, earning degrees from Harvard and a surgical fellowship at the Mayo Clinic in Rochester, Minnesota. That fellowship helped him become a flight surgeon at the Army School of Aviation Medicine (ASAM) in 1937. During his stint at ASAM, Lovelace helped develop a new oxygen delivery mask for pilots flying at high altitudes. To demonstrate the mask's reliability, Lovelace strapped two oxygen bottles to

his legs, donned the mask, and planned to jump out of a bomber at an altitude of 40,200 feet. The air temperature at that altitude was a frigid minus
47 degrees Fahrenheit. His jump would be a static line leap in which his
parachute, connected to the plane, would automatically deploy once the line
reached its maximum length.

The plane took off, and after reaching the desired altitude, Lovelace
stepped out of it. Within seconds, the static cord grew taut. The static line
jerked Lovelace as it deployed the chute, ripping off one of the doctor's
gloves and knocking him unconscious.

Twenty-three minutes later, with barely enough time to brace for landing,
Lovelace awoke and made a successful landing. He was unscathed save for
some frostbite on his ungloved hand. Although he could have been reprimanded for this foolhardy test, he was instead awarded the Distinguished
Flying Cross for his successful demonstration of the mask and for making
a record-breaking skydive simultaneously. It was Dr. Lovelace's first—and
last—parachute jump.[4]

In 1947, Lovelace joined his uncle in New Mexico and established the
Lovelace Foundation for Medicine Education and Research. Soon after, the
man who would become NASA's first administrator, T. Keith Glennan, ap-
pointed Lovelace and Brigadier General Donald Flickinger to head the Special
Committee on Life Sciences, where they were charged with developing and
administering medical and psychological tests for America's first astronauts.

In all, twenty-four applicants were subjected to the grueling tests. They
were poked and prodded in unmentionable places, survived cardiopulmo-
nary stress testing as well as heat stress tests in compact ovens. The list of
evaluation tests seemed endless.[5] Of the twenty-four applicants tested, only
seven made the final cut, and on April 9, 1959, America's Mercury 7 astronauts were introduced to the world. But while the seven were being pum-
meled with questions from the overeager journalists in attendance who were
nearly climbing over one another to ask sometimes insightful and sometimes
silly questions, Dr. Lovelace sat quietly on the stage, aware that there were
other candidates waiting in the wings who were equally qualified to endure
the rigors of spaceflight.

While Dr. Lovelace and General Flickinger were administering the
tests to the Mercury 7, they pondered whether women would make good

astronauts. Lovelace approached a good friend of his, famed pilot Jackie Cochran, to help with the new project, which was informally called the Women in Space program.

Cochran was a steadfast supporter of women in aviation, having herself set several speed and altitude records, won countless aviation races, and become the first woman to break the speed of sound in 1953 and then to reach Mach 2 in 1962. In one year alone, she broke seventy-three records. What Cochran was most noted for, however, was the formation of the Women's Air Force Service Pilots (WASP) program during World War II, which sought women to help in the war effort by testing aircraft and shuttling them across the Atlantic to the front.

Being such a champion for women in aviation, Cochran and her husband Robert were more than happy to fund the fledgling female astronaut testing program when Lovelace approached her about it in 1960. Even before the funding was secured, Flickinger and Lovelace already had in mind who they wanted to be the first woman to join the program—Geraldyn "Jerrie" Cobb.

Jerrie was born in Norman, Oklahoma, in 1931. The young girl had a speech impediment that caused her excruciating embarrassment in elementary school due to classmates who often mocked her. She often told what friends she had that "School is no good. Sometimes it's best to be alone."[6] And that's what she did. Keeping mainly to herself, she found that spending time with horses was much more enjoyable than spending time with people.

Cobb's father Harvey was in the National Guard when World War II broke out. He was fascinated with flying and requested a transfer to the Army Air Corps, but the request was denied due to his age. Even though the military said he was too old to fly, the Air Corps gave him a glimmer of hope when it told him that he could transfer only if he had a commercial pilot's license. Harvey Cobb did not possess the prestigious license but was keen on making sure he got one.

In 1940, Harvey Cobb came home with a spring in his step as he proudly announced to his wife and daughters Jerrie and Carolyn that he had finally earned his license. To celebrate his achievement, Cobb treated himself to a special gift—he purchased a small Taylorcraft airplane.

The two girls were ecstatic and couldn't wait to take their first flight. The day following the announcement, the two Cobb girls skipped school

Jerrie Cobb poses next to a Mercury capsule. Although she never flew in space, Cobb, along with thirteen other women, underwent physical tests similar to those taken by the Mercury astronauts with the belief that she might become an astronaut trainee. *NASA*

and their father took them on their first airplane flight. Jerrie Cobb later recalled the moment: "Even before the old Taylorcraft reached three hundred feet," she said, "I recognized the sky would be my home. For a child that distrusted ordinary, everyday speech, for an adolescent who yearned for the freedom of the fields and winds, for a girl who had learned to be alone—the sky was the answer."[7]

And indeed it was. At age sixteen, Cobb earned her pilot's license. By eighteen, she had her commercial license and became a certified ground instructor with ratings in civil air regulations, meteorology, and engine mechanics. She went on to ferry B-17 bombers around the globe and, for good measure, broke several speed, distance, and altitude records.

The tests that Lovelace and Flickinger developed for the first phase of their program were intense to say the least, but Cobb was ready for them: four-hour eye exams, a full day of x-raying every bone and tooth in her body, pedaling for hours on a specially weighted stationary bike to test respiration that drove her to exhaustion. She spent considerable time inverted on a tilt table to test her blood circulation, had electric pulses shot through her joints to test her reflexes, and swallowed rubber tubes to test stomach acid.

Then there was the most excruciating test of all—the ice water test in which cold, 10 degree water was shot into her ear by a syringe. The frigid water immediately froze the inner ear causing severe vertigo. In an interview with Carol Butler for the NASA Johnson Space Center Oral History program, a member of the next group of women to take that same test, Mary Wallace "Wally" Funk, said that as soon as the water hit her ear, "the room began to spin [uncontrollably]. You were supposed to focus on a single spot. I couldn't lift my arms and would have fallen off the table to the floor if I weren't strapped in."[8]

When all was said and done, Cobb had aced phase one testing. On August 19, 1960, Dr. Lovelace appeared at the Space and Naval Medicine Congress in Stockholm, Sweden, where he announced the results, stating "We are already in a position to say that certain qualities of the female space pilot are preferrable to those of her male counterpart." He then went on to open the floodgates for women to join his program by adding, "There is no question but that women will eventually participate in spaceflight, therefore we must obtain data on them comparable to what we have obtained on men."[9]

To be selected for the Women in Space program, applicants were required to be pilots with at least one thousand hours of flight time, have a commercial rating, and possess a college degree, though the latter was not a firm requirement. Out of all the applications received, twenty-five women were invited to take the second phase of testing. Jackie Cochran was brought in as an adviser, and she and her husband paid for all of the women's expenses.

Nineteen of the invited applicants would test at Lovelace's New Mexico test facility. The women would not test at the same time. Instead, the tests would be administered either individually or the women would be scheduled in pairs. It wasn't until years later that the final group of women who were selected would meet the other candidates.

The oldest applicant was forty-one-year-old Jane Hart, a mother of eight whose husband was Michigan Senator Phillip A. Hart. The youngest was twenty-three-year-old flight instructor Wally Funk.

In addition to the tests mentioned earlier, for the second phase of testing, each of the women was strapped into a centrifuge at Oklahoma State University to test her body's reaction to the force of gravity exerted during a launch. Wally Funk recalls that civilians were not allowed to wear pressure suits, and because of that, the women were only allowed to spin at 3Gs. From her pilot training, she knew that when excessive gravity is placed on the human body, a person's blood flow is forced downward, away from the brain, which could cause the pilot to black out. To avoid this, the blood had to be forced upward, and without a pressurized flight suit, blacking out was highly likely thanks to the dizzying tests the women would endure.

Funk wrote to her mother and asked if she would send her a merry widow. When it was her turn in the centrifuge, Funk wore the tight-fitting girdle beneath her clothing to help equalize the pressure, and she made it through with flying colors.

Three of the subjects—Funk, Jerrie Cobb, and Rhea Hurrle—were also tested in an isolation tank. Men went through a similar isolation test, but the male version of the test was set up in a small, dimly lit or completely dark room where they sat in a chair, having plenty of time to contemplate what was going on. The women, on the other hand, were made to float on their backs in a water tank. The humidity in the room and the water temperature was balanced to match the women's body temperature exactly, which caused them to lose all sensory perception—sight, hearing, smell, and feel. The women recall it being a blissful and tranquil experience. Funk, in fact, set a record by remaining in the tank for ten hours and thirty-five minutes and had to be asked to terminate the test.

The women also took part in ejection seat testing, wherein they were strapped to a seat attached to a vertical rail that shot straight up in the air like a rocket. The effect pushed 12Gs of force on their bodies before they came back down with a thud.

Lastly, the women participated in high altitude evaluations by traveling in a plane to an altitude of more than thirty-nine thousand feet while breathing pure oxygen. Once they reached altitude, they removed the oxygen mask and were asked to write letters and numbers on a notepad. The human body can last only four to six minutes without oxygen before passing out. In Wally Funk's case, she quickly began to black out when she removed the mask. The technicians overseeing the test saw the condition she was in and ordered her to put the mask back on, but she didn't respond. They rushed into the test area of the plane where Funk was seated and placed the oxygen mask back on her face, and just that fast, she was revived. Funk said that during the test without the mask on, she believed that she was acing it and writing perfectly, but when she looked at the paper afterward, all she saw were lines of incomprehensible gibberish.[10]

As the summer of 1961 slid into fall, thirteen of the nineteen women—Cobb, Funk, Hurrle, and Hart, as well as Myrtle Cagle, Janet Dietrich, her twin sister Marion Dietrich, Sarah Gorelick, Jean Hixson, Gene Nora Stumbough, Irene Leverton, Jerri Sloan, and Bernice Steadman—had passed the phase two tests. Informally, the thirteen women called themselves the "first lady astronaut trainees," or FLATs. When the media got word of their achievements, they became known as the Mercury 13.

The Mercury 13 had performed admirably, and they achieved test scores much higher than their male counterparts. Lovelace and Flickinger invited the women to participate in the third and final series of tests at the Naval School of Aviation Medicine in Pensacola, Florida. A few of the women quit their jobs so that they could attend training, but only a few days before their arrival in Florida, the thirteen women were informed via telegram from NASA that because the Lovelace evaluations were privately funded and were not authorized by NASA, the training had been cancelled.

It's true that being a civilian project most likely caused the cancellation of the program, but beneath it all was the nagging stereotype that women were not suited to go into space. Shortly after the hopes of the Mercury 13 were dashed, Linda Halpern, a young schoolgirl who was enthralled by the space program, wrote to President Kennedy to ask how she could become an astronaut. The letter was forwarded to NASA's Office of Public Affairs. The response was cordial but to the point: "while many women are employed in other capacities in the space program . . . we have no present plans to employ women on space flights because of the degree of scientific and flight training, and the physical characteristics which are required."[11]

Another woman whose name has become a footnote to this story and pushed to the back pages of history is Betty Skelton. Born in Pensacola, Skelton would become an outstanding pilot in her own right, having made her first solo flight at the age of twelve when Navy Ensign Kenneth Wright, who was teaching her to fly, let her take control of a Taylorcraft aircraft (her first legal solo was at the age of sixteen).

Skelton went on to break many speed records, and according to the Smithsonian's National Air and Space Museum, she holds more combined aircraft and automotive records than anyone in history.[12] She began her career as an aerobatic flyer in 1946, taking to the air for her first air show at the 1946 Southeastern Air Exposition, sharing the skies with a brand-new flying team that was also just starting out, the Blue Angels. Skelton and the Blues became close friends and she was known as the "Sweetheart of the Blues."[13]

The aviator earned the respect of one of the future Mercury 7 astronauts, Wally Schirra, when Schirra and his squadron at the Pensacola Naval Air Station sat in a hangar in 1948 watching a torrential Gulf Coast downpour. Through the rain, they heard the engine of a tiny biplane and watched as the pilot skillfully landed in the storm. The men watched as the pilot stepped from the plane. Who was he? The pilot removed his helmet and long brown hair poured out. The pilot wasn't a man but a woman—Betty Skelton.[14]

In 1960, *Look Magazine* prepared to run the cover story, "Should a Girl Be First in Space?" and asked Skelton to take the same tests that the Mercury 13 women had. Once again, a woman passed the tests with flying colors, for which Schirra called her "Mercury 7½," but Skelton was realistic about the results, saying that the test was just a media stunt.

Jerrie Cobb and Jane Hart, on the other hand, were not going to stand idly by and let their chances to become astronauts slip away. In a letter to President Kennedy and Vice President Lyndon B. Johnson, the women spelled out the reasons why the Lovelace program needed to continue and asked that it be restarted. Johnson did query NASA Administrator James

Betty Skelton poses in a Mercury spacesuit for a *Look* magazine article on women in space after completing the same training as the Mercury 7 astronauts and Mercury 13 women. *Joe Cuhaj*

Webb about whether women had been turned down for a seat in a capsule simply because they were women, but at the bottom of his letter to Webb, he scribbled the words, "Let's stop this now!"

Hart petitioned and was granted a hearing on sex discrimination with the US House of Representatives Special Subcommittee on the Selection of Astronauts. The hearing convened on July 17, 1962, with Hart and Cobb testifying on behalf of the Mercury 13. The two women laid out a powerful case for including women in the astronaut corps. Hart argued that "for many women, the PTA just is not enough. . . . I don't want to downgrade the feminine role of a wife, mother, and homemaker. . . . But I don't think it unwomanly to be intelligent, to be courageous, to be energetic, to be anxious to contribute to human knowledge."[15]

For the most part, the hearings were cordial and many of the congressmen were sympathetic to the women's aspirations, but the inevitable stereotypes were still present:

> Mr. [Victor] Anfuso [congressman from New York following opening statements by Jane Hart]: Miss Cobb, I think that we can safely say at this time that the whole purpose of space exploration is to someday colonize these other planets and I don't see how we can do that without women. . . . [Laughter] I call on Mrs. Hart.
>
> Mrs. Hart: I would like to say, I couldn't help but notice that you call upon me immediately after you referred to colonizing space.
>
> Mr. Anfuso: That's why I did it. [Laughter][16]

Also testifying were two of the Mercury 7 astronauts, Scott Carpenter and John Glenn, who spelled out the requirements for being a NASA astronaut, which included having engineering degrees and being a military jet test pilot, two qualifications that would not be afforded to women for decades.

Glenn put the exclamation point on the proceedings when he testified, "men go off and fight the wars and fly the airplanes and come back and help design and build and test them. The fact that women are not in this field is a fact of our social order."[17]

The fate of the Mercury 13 was sealed, and the program officially ended.

During the hearing, Jerrie Cobb was asked by Representative Anfuso if she believed that the Soviet Union was planning to launch a woman into space. "Yes," she replied. "It is a known fact that they have women pilots in their armed forces."

"Do you feel it is essential to have been a test pilot before you qualify as an astronaut?" another congressman asked.

"I personally feel it is not essential at all," she replied.

Cobb was correct. The Soviet Union was planning to send a woman into space and the fact that she was not a pilot was not a disqualifying factor.

In late 1962, a delegation of Soviet engineers and cosmonauts (including Yuri Gagarin) visited the United States as part of a worldwide promotional tour. During the visit, they dined with President Kennedy and met with their American counterparts. After the fanfare of their visit had faded to memory and the delegation headed home, they took with them more than just memories and souvenirs. They left with the impression that the Americans were preparing to launch a woman into space.

Of course, the Soviets could not be upstaged, and a plan was quickly put in place to beat the Americans to the draw. The push was initiated by the head of the cosmonaut training program, General Nikolai Kamanin, who had trained the country's first three cosmonauts—Gagarin, Gherman Titov, and Alexei Leonov. Kamanin recruited some serious support for the effort, including the backing of the man who pioneered manned spaceflight in the Soviet Union, Sergey Korolev.

The requirements to win a seat aboard a Vostok capsule were simple: women must be younger than thirty years old, weigh less than 155 pounds, and be shorter than five-and-a-half feet tall. Unlike American astronauts, a college degree and flight experience were optional. There was one additional requirement that was nonnegotiable. Each applicant must have skydiving experience. The reason for this was that the Vostok spacecraft wasn't designed to land with the cosmonaut inside of it. Instead, they would eject from the spacecraft high above the ground and parachute to a safe landing separate from the capsule.

As soon as the requirements were laid out and the ink had dried, invitations were sent out to female members of local skydiving clubs across the country. Most of the invitations were clandestine in nature. For example, Muscovite Valentina Ponomaryova was a twenty-eight-year-old staffer at the Department of Applied Mathematics at Steklov Mathematical Institute. At a New Year's Eve office party, a colleague asked her to dance. As they swayed to the music, he whispered in her ear, "Would you like to fly higher than any pilot?"

Ponomaryova thought that the comment was obviously a joke and laughed it off, but after being approached several more times by her colleague, she decided to take a chance and submitted an application to be a cosmonaut.

The application made it to the president of the USSR Academy of Science, Mstislav Keldysh, who invited her to his office for an interview. His first question to her was, "Why do you like to fly?" She thought about it for a moment and replied, "I don't know."

Keldysh looked at the young woman and replied, "That's right. We can never know why we like to fly."[18]

Despite an objection by Gagarin, who said, "We cannot put the life of a mother at risk by sending her to space" (Ponomaryova had one child), her application was accepted.

In all, eight hundred women applied to the Soviet space program, fifty-eight of those were formally considered, and only twenty-three made the final cut to take part in advanced training and medical evaluations. One of the fortunate few was twenty-six-year-old Valentina Tereshkova.

Tereshkova was born into a working-class family on March 6, 1937, 170 miles northeast of Moscow in the town of Maslennikovo. As a young teen, she worked at a textile factory where she quickly rose through the ranks of the Communist Party and was elected secretary of the youth organization the Komsomol Committee. During these formative years, her love of skydiving began. She joined a local parachuting club and took her first jump in 1959 at the age of twenty-two. Her loyalty to the Communist Party and the 126 jumps she had made by 1961 made her a prime candidate for the program.

Eventually, the final twenty-three candidates were cut to five. They became known as the Space Squad: Zhanna Yorkina, Irina Solovyova, Tatyana Kuznetsova, Valentina Ponomaryova, and Valentina Tereshkova. The women were sent to Star City, a former Soviet air force base hidden away in the forest just northeast of Moscow, which had been converted into the official cosmonaut training center where the Space Squad was first introduced to the male cosmonauts who were also in training.

"The guys," Ponomaryova said, "treated us well . . . but they were not happy when we five girls first showed up in Star City."[19]

Medical evaluations began in late 1962 and were the same as those administered to the men—a long series of x-rays and brain studies, advanced cardiovascular and blood screenings, and the dreaded centrifuge.

The women were always confined to the base and never allowed to leave—that is, except for three who were allowed to venture out from time to time because they lived in Moscow, leaving Yorkina and Tereshkova behind. One day, the pair became bored and asked their supervisor if they could spend the day in the city. The man asked, "What for? What do you want to buy?"

Tereshkova lost her cool and shouted, "*Knickers*! That's what we want to buy!"

The two were allowed to leave.

Meanwhile, the Soviet Union scored another first by putting two manned spacecraft in orbit at the same time. Upon returning to Earth, the Kremlin made the decision to replicate the mission, but this time a serious debate ensued about who should pilot the twin Vostok spacecraft: Should they fly the mission again with two men? Should it be a man and a woman? How about two women?

The final decision was made by a group of politicians that included Soviet Prime Minister Nikita Khrushchev. It was said that Khrushchev was looking for a "Gagarin in a skirt."[20]

It was decided that a man and a woman would fly the two spacecraft. The male pilot would be Valery Fyodorovich Bykovsky. His mission in Vostok 5 would not only rendezvous with another Vostok spacecraft as had been done just months before, but also set a new space duration record. The choice of the first woman in space wasn't as cut-and-dried. Everyone assumed the choice would be Ponomaryova, but mission planners worried that her "moral values were not stable enough." That and an almost debilitating landing during an earlier parachute jump completely knocked her out of the running.

The choice was made—it would be Tereshkova.[21]

Bykovsky's launch was plagued with delays. It was originally scheduled for June 7, 1963, but it was postponed due to high winds, which violated launch constraints. Two days later, the Vostok rocket was erected on the pad, but the launch was postponed again due to an expected increase in solar flares. Even though the Vostok was protected from radiation, scientists had no clue as to what increased radiation would do to the craft or passenger.

The launch was further delayed on June 14. With only one hour before liftoff, engineers realized that the ejection seat was not cocked and loaded. This would have been a fatal error, since that was the only way the cosmonaut could land—shooting out of the capsule in his seat to parachute to the ground sans capsule. The capsule had to be reopened.

Once the ejection seat was ready, the countdown continued until, at T-minus five minutes, one of the gyroscopes on the rocket failed. However, engineers felt confident enough that they could launch Bykovsky safely and resumed the count. Finally, at 14:58 Moscow time, Vostok 5 was launched.

Aboard Vostok 6, Tereshkova had a much easier time with a nearly flawless launch. With the code name Chayka (seagull), Tereshkova took wing right on time, rocketing into the skies over the Baikonur launch complex on June 16, shouting joyfully, "Hey sky, take off your hat, I'm on my way!"

The mission itself was a huge success, with Vostok 5 and Vostok 6 orbiting the Earth simultaneously, getting as close as three miles apart, which allowed them to communicate with one another. Thirty minutes after communications were confirmed, Moscow began blaring the news to the world of yet another Soviet space first—a woman was in space.

Everything was going well for Tereshkova, with only a few minor exceptions such as a nagging pain in her right shin that became "very disturbing by the third day," her inability to complete biological experiments because she couldn't reach the equipment, and a dislike for the food.

"The bread was too dry," she said in her postflight debriefing. "I didn't eat it. I mainly ate the black bread and tubed onions. I threw up once, but it was due to the food and not due to vestibular problems."[22]

Later reports suggested a different story—she had lost her appetite, vomited more than once, was losing weight, and as ground controllers radioed the craft to discuss her upcoming landing, there was no response. Turning on the cabin's television camera, they found her asleep and flipped on the cabin lights to wake her.

The cosmonaut continually told mission control that she was fine, but those on the ground believed that fatigue was setting in and began complaining that they were not getting clear responses from the world's first woman space farer.

As reentry time neared, the Vostok's automatic reentry software began to fail, causing the spacecraft to begin a slight tumble, pushing the craft out of position for the required braking burn that would start her descent back to Earth. This would require Tereshkova to do a manual burn of thrusters to position the craft correctly. When she failed to make the burn on her first opportunity, Korolev became livid, taking it out on the Russian equivalent of the Americans' capsule communicators, shouting at them that they "need to teach her how to operate the spacecraft."[23]

On Vostok 5, ground control radioed to Bykovsky that due to insufficient boost during launch, his orbit was decaying and he would have to make an early landing. Both Bykovsky and Tereshkova were then sent the command to begin braking and reentry. Bykovsky replied that he was ready. There was no reply from Tereshkova.

Unbeknownst to mission control due to the loss of radio contact with the cosmonaut, Tereshkova's craft had already made the burn and was beginning to deorbit, heading for a landing near the Kazakhstan-Mongolia border. Although Tereshkova's debriefing told a different story, many believed that she was partially unconscious after ejecting from the capsule and parachuting to the ground. Upon landing, she broke all protocols by not waiting for medical teams to arrive. With the aid of local villagers who had arrived on the scene first, she stepped out of her flight suit and accepted offerings of food. In exchange, she handed out her remaining space rations.

It would take almost seven hours for mission control to receive word that Chayka had landed safely. When the news finally reached Korolev that the pilot was safe and that the mission had ended successfully, he remarked, "So now I'll have to agree one more time to launch a broad."[24]

Despite breaking the rules, Tereshkova became a national hero, winning the official title "Hero of the Soviet Union" and receiving the Soviet Union's Order of Lenin and the Gold Star for bravery and service. She also received the United Nations Gold Medal of Peace.

Once again, the Soviet Union had chalked up another space first, and as news of her journey spread around the globe, Tereshkova became a worldwide hero, not only for her accomplishment, but also for opening a door for women. Author and former congresswoman Clare Booth Luce wrote in an article for *Life* magazine that Tereshkova had "orbited over the sex barrier."

But just as quickly as that door opened, it slammed shut again. It would take the Soviet Union another twenty years to launch their second woman into space, Svetlana Savitskaya, aboard the Soyuz T-7 mission in 1982. Two years later, she became the first woman to walk in space.

It took the Americans much longer to launch their first woman into space. NASA's class of 1978 saw thirty-five astronauts selected to fly on the space shuttle. Of those, six were women—Shannon Lucid, Margaret Rhea Seddon, Kathryn Sullivan, Judith Resnick, Anna Fisher, and Sally Ride. Ride would be the first into space aboard the space shuttle *Challenger* on mission

STS-7 in 1983, when she used the robotic arm to launch the first shuttle pallet satellite (SPAS-01). Upon selection, Ride thanked actress Nichelle Nichols, who portrayed Lieutenant Uhura in the original episodes of the television show *Star Trek* for "recruiting" her into NASA.

Nichols had contracted with NASA in 1977 to create a recruitment film for the new breed of astronaut.[25] "If it wasn't for you," Ride told the actress, "I might not be here."

All the other women in the class would fly at least one mission. Judy Resnick was an electrical and biomedical engineer who flew on the maiden voyage of *Discovery* on STS-41D as a mission specialist. Before the flight, a Jewish magazine asked her to do an interview. She declined, telling her father, "I don't want to be a Jewish astronaut. I don't want to be a Jewish woman astronaut. I just want to be an astronaut, period."[26]

Her promising, pioneering career in space was tragically cut short on January 28, 1986, when the boosters launching *Challenger* exploded, killing the crew of STS-51L.

Kathryn Sullivan made history as the first American woman to walk in space in 1984, and in 2014 she was confirmed by the US Senate as the tenth director of the National Oceanic and Atmospheric Administration (NOAA). Sullivan made history once again in 2020 when she visited a place where no other woman had gone before, the deepest depth in any of Earth's oceans, the Challenger Deep. Sullivan was sixty-eight years old when she boarded the submersible to make the seven-mile journey below the surface of the Pacific Ocean.

One of the women who was later selected for the astronaut corps, Eileen Collins, paid homage to the Mercury 13 women when she invited the surviving members of the group to witness the launch of shuttle mission STS-63 on which Collins became the first woman to command and pilot a space shuttle.

In written testimony to the US Senate in 2012, Mae Jemison, the first African American woman in space, neatly summed up the advances that women were making in space exploration. "Today, although women still represent a minority of the astronaut program, our space program is more inclusive. And, as a nation, we accept women in space as a routine occurrence."[27]

Even though it took more than twenty years for the United States to send a woman into space, women were still filling important roles back on Earth for NASA. We have all heard the story of the African American "computers" from the hit movie and best-selling book, *Hidden Figures*, but there were other women who also have stood out through the years in NASA history.

Up until the flight of Apollo 11 and mankind's first landing on the moon, women were not allowed into the firing room at the Kennedy Space Center. That all changed when JoAnn Morgan was put in charge of the instrumentation console and communications and broadcast systems for Apollo 11. The title of "first woman to serve as a NASA engineer" went to Frances "Poppy" Northcutt. She was the only woman working at mission control during Apollo 8's historic first manned flight around the moon in 1968. She was also one of the engineers who helped "MacGyver" a solution that brought the Apollo 13 crew home from the moon after an oxygen tank exploded and crippled the spacecraft.

Margaret Hamilton is credited with coining the term "software engineering." As a NASA contractor on loan from MIT, Margaret helped develop the software for Apollo 11, winning the Presidential Medal of Freedom in 2013.

Despite a slow start, the history of women in space has been accelerating. The most exciting advancement for women in space came when NASA announced the eighteen astronauts to train for the United States' return to the moon. Of the eighteen trainees, nine are women, and one of them will be the first woman to set foot on the lunar surface. The project is appropriately named Artemis, the goddess of the moon and the twin sister of the Greek god Apollo. The women selected include Stephanie Wilson, Jessica Watkins, Kate Rubins, Jasmin Moghbeli, Jessica Meir, Anne McClain, Nicole A. Mann, Christina H. Koch, and Kayla Barron.

In another first for women in space history, NASA has introduced its first female launch director, Charlie Blackwell-Thompson, who will guide the launch of the largest rocket ever built, the Space Launch System, as it sends the women of Artemis to the moon.

Before we leave these tales of women who are pioneering space travel, we need to return to the original Mercury 13. As I finished writing this book in the summer of 2021, an extraordinary tweet was sent by Amazon founder Jeff Bezos. As is the case with billionaire Sir Richard Branson and his Virgin Galactic project, Bezos wants space flight to be more accessible to the public and created a new commercial enterprise, Blue Origin, to do just that.

Charlie Blackwell-Thompson, chief NASA test director, works at a console in Firing Room 4 during the countdown for STS-133. *NASA/Kim Shiflett*

The first flight of the company's New Shepard rocket, which was named for America's first astronaut, was set for the morning of July 20 from a desert launch site in west Texas. After fifteen successful unmanned launches, New Shepard took four passengers on a ten-minute suborbital ride. The passengers included Bezos, his brother Mark, eighteen-year-old Oliver Daeman of the Netherlands (the original third passenger gave up his seat after paying $28 million for the ride), and one more very special guest.

In the video that was attached to the Bezo's tweet, he is seen talking with Mercury 13 trainee Wally Funk. You can see the eighty-two-year-old's face light up as Bezos describes the flight. Later, after describing the flight plan, Bezos tells the veteran pilot, "[that is your] very first flight." Before Bezos can complete his sentence, Funk grabs the man and pulls him in for a long, hard bear hug. Sixty years after being denied the chance to make history and live her dream, Wally Funk's long countdown was to finally reach liftoff.

"In 1961," Bezos wrote in the tweet, "Wally Funk was at the top of her class as part of the 'Mercury 13' Woman in Space program. Despite completing their training, the program was cancelled, and none of the thirteen flew.

"It's time. Welcome to the crew, Wally. We're excited to have you fly with us on July 20 as our honored guest."[28]

And so it was that on July 20, 2021, at 9:00 a.m., Funk rocketed skyward from the desert to an altitude of sixty-six miles for a ten-minute suborbital flight. The veteran aviator was giddy as she tumbled weightless for two minutes with her crewmates.

Not only was the flight a culmination of the dream of the Mercury 13, but it also put the aviator in the record books—the oldest person to fly in space, beating John Glenn's record by five years. The record was short-lived, however; only three months later, *Star Trek* star William Shatner rode the same rocket into space at age ninety.

More important than the age record, Wally finally had earned her astronaut wings.

3

IN MEMORY OF LAIKA

Before a human could safely fly into space, a myriad of questions had to be answered: What are the dangers of cosmic radiation? How will the body handle weightlessness? Can a human handle the forces exerted on it during launch? How can we find the answers to these questions without sacrificing human life?

The answer to the last question was simple for engineers—launch animals into space.

To say that mankind's voyages into space would not be possible were it not for test flights conducted with a multitude of animals is an understatement. But these unwitting heroes of the space age—frightened and alone in a tight-fitting sealed capsule—often paid the ultimate price in the name of science and exploration.

The most famous animal astronaut was the Soviet Union's dog Laika, but before Laika's ill-fated journey, the United States had already been sending animals to the edge of space. It all began with Project Blossom.

As the German rocket scientists that surrendered to the Americans at the end of World War II began settling into their new assignments in the deserts of Texas and New Mexico, the team was reunited with the parts from their captured V-2 rockets and began the process of reassembling them for test flights. Along with Cambridge Research Laboratories, the Air Force modified several of the rockets, increasing the body length by sixty-five inches and configuring the warhead section so that it could be used to hurl scientific instrumentation and biological specimens to the edge of space.

Early test flights of the modified V-2 rockets, complete with fruit flies on-board as their initial payload, exploded only minutes after launch. Still, Project Blossom forged ahead as engineers prepared to begin experiments using rhesus monkeys as their test subjects. Monkeys, they believed, were docile and able to learn and perform complex tasks. And, of course, they were as close to a human subject as they could get without actually flying a human.

The capsules that were designed for the flights were a snug fit to say the least—a cylindrical enclosure measuring three feet long and twelve inches in diameter, just big enough to secure a nine-pound monkey and cram scientific equipment around it in the nose cone. The anesthetized monkey was placed in a foam seat attached to a metal rack. A net suit was then placed around the monkey before he was strapped in. The entire unit—monkey and all—was slid into the small capsule, sealed inside, and the cabin pressurized.

The pressurization of the capsule proved to be a thorn in engineer's side. When fully pressurized, the walls of the capsule bowed out much like a balloon. Metal reinforcements were added to prevent this. The following test revealed numerous holes in the capsule, which allowed oxygen to shoot out like geysers. Technicians made the decision to re-weld all of the seams and, for good measure, added a healthy slathering of caulking around those welds. Consider it a high-tech duct tape fix.

While the capsules were being readied, several test subjects were training at MacDill Air Force Base in Tampa, Florida. During training, one of the monkeys escaped from his handler and was on the lam. Military police and base personnel scrambled to find the missing monkey, but he was elusive. Three weeks after his great escape, the project manager, First Lieutenant David Simons, received a message from the Tampa Police Department stating that they had caught a monkey and asked if it belonged to the officer. As Simons later described:

> It seemed this monkey was tired of the Air Force base and wandered off into town a few miles away. It stuck its nose into a lady's kitchen one morning and began to snoop around. The lady happened to be a meticulous housekeeper, especially in the kitchen. Her concept of neatness naturally did not include having a monkey crawling around among her saucers and teacups. So, she made the mistake of trying to remove the monkey from her cupboard by force. The monkey took exception to her attack. He started throwing teacups

and saucers in her direction. She and the monkey then began running around and around her kitchen, until it became rather the worse for wear![1]

The incident was settled in court with the woman winning several hundred dollars for emotional damages, broken dishes, and the "general besmirchment of her clean kitchen."

With the monkey back in its proper place and all of the kinks in capsule design having been ironed out, another V-2 was prepared for launch and, for the first time, would have a monkey as a passenger. The launch occurred on June 11, 1948. Engineers strapped a rhesus monkey named Albert into his seat, attached sensors to his body, then loaded him into the V-2. According to Simons, Albert's capsule was so small that the monkey's head had to be "placed into a cramped, forward position with the neck acutely flexed."

After Albert was sealed in and the countdown clock started, the mission team settled in to track Albert's vital signs during the flight. Not long after, the team discovered that they were not picking up any readings from Albert's sensors. Engineers began troubleshooting the issue and realized that there were only two possible answers: either the sensors were defective or Albert had died. This late in the countdown, the engineers did not want to open the capsule and check on its passenger and decided to press on with the launch.

On the launch pad, the V-2 emitted a short-lived puff of white smoke from its tail section. Within seconds the engine ignited, sending a pool of orange flame radiating across the ground around its base. The rocket's engine burst fully to life and the black-and-white rocket shot off the launch pad, its flame cutting the desert sky like a knife.

At an altitude of thirty-eight and a half miles, the engine prematurely shut down. An engine valve had failed. The rocket began plunging down. Upon reaching 25,000 feet, the nose cone separated and continued its plummet toward Earth. The parachutes deployed but couldn't flare due to the thin atmosphere at that altitude. By the time the nose cone reached the thicker lower atmosphere and the parachutes grasped enough air to unfurl, the impact ripped the chutes apart, the nose cone carrying its tiny passenger hit the desert floor, tumbling end over end. If Albert wasn't dead at launch, he surely died on impact.

One year later, a second V-2 was prepared with another passenger, a second rhesus monkey named Albert II. Sensors indicated that Albert II survived the punishing forces exerted on his body during launch, reaching an altitude of eighty-three miles, making him the first animal or human to reach space (an altitude of one hundred kilometers, or slightly more than sixty-two miles is considered suborbital and the edge of space).

Albert II did not get the chance to relish in the glory of his accomplishment, however. As with his predecessor, the parachute once again failed, and the capsule plunged to earth, killing its occupant.

Two more attempts were made to complete a successful flight with rhesus monkeys. Albert III died in September when his rocket exploded ten seconds after launch. Albert IV had a successful launch, reaching an altitude of seventy-nine miles, becoming the second living creature to make it into space, but for the third time, the parachute failed, and Albert IV was killed on impact.

On the other side of the world, the Soviets were keeping tabs on the US efforts and initiated their own program to study the effects of spaceflight on living beings. Soviet rocket expert Sergei Korolev and biomedical researcher Vladimir Yazdovsky agreed that their program would rely mainly on dogs. Their thinking was that dogs, especially female dogs, were calmer and would be less fidgety than monkeys in the close quarters of a pressurized cabin. Female dogs also had better control of their bowel movements. Overriding it all was the fact that dogs could be found literally everywhere on the streets of Moscow. Strays filled the streets, so there was an ample supply of subjects, plus it was assumed that a stray could better handle the stresses put on it during launch and recovery after living a tough life on the street.

For initial test flights, the scientists made the decision to send two dogs into suborbital space at the same time so that they could obtain more accurate results and make physiological comparisons between the two before and after the flight. To train the dogs to sit calmly in a cramped, confined capsule for hours or even days on end, the dog's trainers put them in boxes. As training progressed, they decreased the size of the boxes, eventually keeping the pups inside for as long as twenty days at a time. Additionally, dogs were subjected to harrowing rides on a centrifuge, were forced to stand for hours at a time, and learned to be comfortable wearing space suits.[2]

The first flight in the Soviet program came on July 22, 1951, when the dogs Dezik and Tsygan were launched to the outer edge of space. Both dogs returned safely to Earth. Dezik also made the follow-up flight in September, along with her copilot Lisa. Unlike the previous flight, it did not have a happy ending: both dogs perished during the flight. It was reported that Korolev was devastated over the loss.

One of the "copilots" for the next flight, a dog name Smelaya, apparently didn't understand that her name means "brave." She knew something was fishy, and during a walk with her trainer the day before the launch, she bolted and hightailed it away. After searching late into the night, the trainer returned to the kennel to prepare a backup dog for the mission, all the while fearing that Smelaya might be eaten by wolves.[3]

As he readied the stand-in for the flight, who should show up but Smelaya herself. With only a short time before launch, the team prepared Smelaya and her partner Malyshka, and the dog duo roared to the fringes of space and were recovered safely. The Soviets chalked up another successful suborbital flight.

Between 1951 and 1966, the Soviet Union launched seventy-one dogs into space (many of them flew twice). Seventeen did not survive. Two of the more notable flights include the escape of another runaway and a flight that bridged the gap between two powerful world leaders.

The first flight of note was the sixth of these early Soviet tests. Much like Smelaya before her, Bobik, one of the dogs trained for the mission, also had a bad feeling about what was in store for her. Maybe Smelaya warned Bobik about what the future held, so like her predecessor, Bobik ran away. Unlike Smelaya, Bobik never returned. The scientists had to tap a backup dog for the flight. The dog chosen was named ZIB by engineers, an acronym that, when translated from Russian, stood for "substitute for missing dog Bobik." ZIB and his companion crew member were successfully launched and returned to Earth on September 3, 1951.

Despite tensions between the United States and the Soviet Union in the early 1960s, another notable flight proved that the two most powerful leaders in the world shared a soft side for dogs.

It began on October 19, 1960, as a rabbit, two rats, a container of fruit flies, assorted plants, and two more Soviet dogs, Belka and Strelka, were launched into orbit. All indications suggested that the canine pair had successfully arrived in orbit, but scientists on the ground began to worry. All telemetry from the pups' sensors were reading normal, but something was wrong. The images that the capsule's video camera was sending back to Earth showed two lifeless dogs. But how could that be? All vital signs were normal.

Scientists watching the video were startled when they saw Belka stir and then begin vomiting. That animated both dogs, proving that they were indeed alive.

It wasn't long before the retro-rockets fired and, after seventeen successful orbits, the newest Soviet "cosmonauts" landed safely. The Soviet Union had flown yet another successful mission and again the world was abuzz with excitement. The dogs became global celebrities, appearing in newsreels and magazines. Commemorative stamps were issued. A headline in the *New York Times* read, "2 Satellite Dogs Shown in Moscow," with the lead "Belka and Strelka Pose for Cameramen but Refuse to Bark for Radio Reporters." The article went on to say that the two dogs appeared perfectly healthy after their flight, even scampering about, playing with the newsmen.

Eight months later, President John F. Kennedy found himself in a tense meeting with Soviet Prime Minister Nikita Khrushchev. It wasn't clear whether the president or his wife initiated the conversation about the space dog Strelka, but the tension was broken. Khruschev replied that the space traveler recently had a litter of puppies. Jackie Kennedy responded, "You must send one of the puppies to me."

Several weeks later, a furry white puppy arrived at the White House complete with a Russian passport. Its name was Pushinka, which means, appropriately enough, "fluffy."

The family loved the new addition. In an interview with Larry King, the late John F. Kennedy Jr. told the reporter how he and his sister Caroline taught the pup to climb the ladder of the slide in the playground behind the White House and slide down.[4]

Pushinka and the Kennedys' other dog, Charlie, went on to have four puppies—Butterfly, White Tips, Blackie, and Streaker. The president nicknamed them "pupniks." A pair of the puppies were given to two children who had written to the First Lady and offered to look after them for her.[5] Within fifteen months, the cute and friendly offering of détente between the two countries would be forgotten as the world woke up to the threat

Kennedy family dog, Pushinka, slides down the treehouse slide on the South Lawn of the White House, Washington, D.C. *Cecil Stoughton. White House Photographs. John F. Kennedy Presidential Library and Museum, Boston*

of a nuclear annihilation as Kennedy and Khrushchev faced off during the Cuban Missile Crisis.

The United States finally completed a successful flight in the Project Blossom series only five days after ZIB's flight. Another rhesus monkey by the name of Yorick took the ride. Yorick was originally named Albert V after his predecessors, but after the previous Albert flights, it was decided his name should be changed. That might have been a stroke of luck for Yorick. He was launched from Holloman Air Force Base in New Mexico on September 20, 1951, attaining a height of 44.7 miles, just short of suborbital altitude, but he returned safely and was recovered in the New Mexico desert.

Yorick was the first American monkey-naut to survive a rocket flight but not the last. On May 22, 1952, two Philippine long-tailed macaques—Patricia and Mike—flew to thirty-six miles and survived the punishing launch that exceeded two thousand miles per hour. Once again, the flight was short of touching the boundaries of space but both monkeys returned safely. Patricia and Mike both died of natural causes, Patricia in 1954 and Mike thirteen years later in 1967.

When it came to orbiting Earth, for animals, as with humans, there was a first. That fateful and tragic honor would go to another stray Moscow dog named Laika.

As mentioned earlier, the Soviet Union had already successfully launched dogs to the fringes of space and returned them safely with a series of suborbital flights. Combined with the successful orbiting of Earth's first satellite, Sputnik 1 in October 1957, the Soviets set their sights firmly on a grander prize—placing the first live animal in Earth orbit, which would then open the door for the first human.

The fortieth anniversary of the Bolshevik Revolution was fast approaching, and Soviet prime minister Nikita Khrushchev wanted to celebrate in a big way. He leaned on his rocket team to make it happen, giving engineers only a scant few weeks to design and test a capsule to house a dog.

What they devised was a small, pressurized compartment comparable to your standard pet carrier or kennel. They built a fan into the cabin to keep the temperature around 60 degrees. Due to weight limitations, the dog would be fed only once during flight[6] and would have enough oxygen for only seven days in orbit before it suffocated. Without a reentry system, the capsule would continue to orbit Earth with its lifeless passenger for an additional five months before it would burn up in Earth's atmosphere. The mission was a guaranteed death sentence for its passenger.

Two dogs were selected for the mission. The primary test subject's name was Kudryavka, meaning "little curly" in English. She was introduced to the world during an interview over Radio Moscow where she barked for listeners. The public was enamored with the pup and began calling her Laika (Russian for "barker"). The name stuck.

This calm but friendly spitz-husky mix was subjected to the same training as previous Soviet space dogs and her handlers grew quite fond of her. It was reported that prior to the flight, despite the rule that Laika would have only one meal, one handler gave her extra food. Another took Laika home the night before the launch to "do something nice for her."

In the dark of night on November 3, 1957, Laika was loaded into Sputnik 2 and at 10:59 p.m. Moscow time, the R-7 rocket with its four cumbersome breakaway engines roared to life. The stocky rocket lit up the night sky as it slowly rose from its launch pad. Inside, Laika's heartrate rocketed, racing more than three times its normal rate and her respiratory rate quadrupled.

Weeks after the flight, the Soviet Union reported that she was gasping for air as the rocket raced skyward but her breathing returned to normal once in orbit.[7]

Indeed, all indications showed that Laika had successfully made it into Earth's orbit—the first animal to do so—but due to a malfunction that resulted in the loss of the spacecraft's heat shield, the temperature inside the capsule began to rise despite the built-in fan. The temperature climbed to more than 90 degrees, and after seven hours, Laika was dead.

Sputnik 2 continued orbiting the Earth for five more months with the lifeless body of Laika onboard before the mission ended with a fiery reentry into the Earth's atmosphere.

For the next two weeks, the Soviets told the world that Laika was alive and well, proving that they were ready for the next step of sending a human into orbit. Eventually, however, the truth came out about Laika's fate, but the revelation escaped much of the world's attention. The public was more interested in and consumed with the impressive scientific achievement the Soviets had pulled off than the cruel death of a stray dog.

A few organizations attempted to draw attention to the use and mistreatment of animals for the sake of science. The National Canine Defense League called for a worldwide moment of silence and the Royal Society for the Prevention of Cruelty to Animals (RSPCA) summoned the public to protest Soviet embassies around the world.[8] The Soviets simply brushed protests off as propaganda stunts.

Memorials for Laika popped up around the world. Just two weeks after Laika had died alone in space, commemorative stamps were issued by various countries. The Japanese government created porcelain statues to honor Laika during the Buddhist Year of the Dog.[9] Six months later, the French erected a marble statue of her at the grand Villepine Animal Cemetery just north of Paris.

The quest to learn about the effects of spaceflight continued in the United States, but results were mixed. In 1958, a test of capsule reentry systems was launched from Cape Canaveral with a mouse onboard as part of the Mouse in Able project (MIA). Minutes after launch, the rocket experienced a malfunction and was destroyed. In June, the mouse Laska survived sixty times the Earth's gravity during launch and forty-five minutes of weightlessness before perishing. The final test of the series one year later included fourteen mice onboard, but the rocket had to be destroyed just after launch.[10]

The American team returned to using monkeys, selecting a squirrel monkey named Gordo to take the next flight. Gordo was launched to an altitude of six hundred miles. His capsule splashed down in the Atlantic Ocean but the floatation system malfunctioned, causing the capsule to sink before he was able to be recovered.

Though tragic, the United States learned many lessons from these flights, which greatly improved the chances of survival for mice, monkeys, and humans atop a giant firecracker. The next American attempt—the flight of Able and Baker—proved that.

The plan was for the rhesus monkey to tap on a telegraph key during the flight. As it turned out, the monkey originally selected was born in India, where monkeys are considered sacred and prohibited from experimentation. To avoid an international incident, NASA switched the monkey for another rhesus by the name of Able.[11]

Miss Baker, on the other hand, was an American-born, eleven-ounce squirrel monkey. For the launch, both monkeys were secured to a foam couch that would absorb the shocks and g-forces during launch and recovery. Miss Baker's small couch would be slid into a polyurethane cradle.

The pair rocketed into the sky atop a Jupiter rocket on May 28, 1959, reaching an altitude of three hundred miles before safely splashing down in the Atlantic. Able died four days later due to a reaction to an anesthetic that was sprayed into her cage as researchers were preparing her for surgery to remove a sensor that had become infected.

Now taxidermized, Able is an exhibit at the Smithsonian Air and Space Museum in Washington, D.C., sitting on the couch that made her famous, gazing toward the heavens.

Miss Baker, on the other hand, lived a long and happy life greeting visitors to the US Space and Rocket Center in Huntsville, Alabama, where she passed away at the age of twenty-seven, becoming the oldest squirrel monkey living in captivity in history. She was interred at the center near its entrance where visitors pay homage to her by placing bananas on top of her granite marker.

Following World War II, our once fragile alliance with the Soviet Union was put to the test. With the Soviets developing nuclear bombs and missiles that could carry them halfway around the world (Soviet Premier Nikita Khrushchev proudly boasted that his country was "making missiles like sausages"[12]), the United States needed a means of keeping an eye on what the Soviets were up to. Their answer was the U-2 spy plane.

Developed by the Lockheed Corporation under contracts with the US Air Force and the CIA, the U-2 resembled a sail plane. Its lightweight design and elongated wings allowed the aircraft to fly to an altitude of seventy thousand feet, travel more than three thousand miles, and carry more than seven hundred pounds of the most sophisticated photographic gear available at the time.[13]

As the plane took its first covert flight over the Soviet Union in July 1956, the United States already knew the plane's limitations, mainly that it would be detected eventually and possibly shot down. They knew that the country would need to develop a satellite that could orbit the Earth completely undetected to continue the reconnaissance program. Two companies, Lockheed and Boeing, were tasked with developing a new, two-stage rocket to loft the proposed satellite into orbit, the Thor-Agena, and Project Corona was born.[14]

Before launching satellites into space, the rocket, which engineers named Discoverer, would have to go through a rigorous series of test flights. It was decided that Discoverer 3 would include a biological payload. Its mission: to observe the effects of hair bleaching from cosmic rays on a living organism. Four black mice from the Holloman Air Base Aeromedical Field Laboratory would be the test subjects.

For the flight (which would launch from Vandenburg Air Force Base in California), each mouse would be sequestered in a small, pressurized cage. Three days' worth of food consisting of peanuts, oatmeal, gelatin, orange juice, and water would be sent along. Each mouse would have a small radio strapped to its back that would monitor heart rate, respiration, and muscular activity.[15]

The launch was plagued with problems. The first launch attempt was aborted because engineers had lost the signals from the miniature telemetry units strapped to the mice. At first, they thought the mice had fallen asleep, so they used a cherry picker to elevate a technician to the nose cone. The technician pounded on the nose cone and even shouted and whistled at it in an attempt to wake the mice. The ground controllers had the capsule opened. What they found was that all of the mice had died. When the cages were constructed, corners had been sprayed with a plastic coating, krylon, to cover any sharp edges. The mice had eaten the coating and died.

The second launch was scheduled for June 3, 1959. As the countdown commenced, a warning sounded. Sensors indicated that the humidity in the capsule had reached 100 percent. Technicians rushed to open the nose cone only to find that the mice were alive and well. As it turned out, when the rocket was in a vertical position the cages were positioned directly above the humidity sensor. When the mice urinated, it covered the sensor, setting off the alarm.

After a three-hour delay, the cabin finally dried out and the rocket was successfully launched. The engine of the second-stage Agena fired as predicted, but a malfunction caused it to deplete its fuel far too fast. The rocket with its passengers never made it to orbit and plummeted into the Pacific Ocean.[16]

That second flight was the last straw for the public, which was outraged, comparing these animal experiments to the ill-fated journey of Laika. The head of the British Society against Cruel Sports lodged a formal complaint to the US Ambassador and the public demanded that missions involving animals be halted until launch and recovery techniques were perfected.

It was the first—and last—Discoverer and Project Corona mission to fly a biological payload, but not the last animal to blast off into space aboard a US rocket.

As mentioned earlier, in the early 1950s, the US Air Force established an aeronautical research facility in New Mexico—Holloman Air Force Base—and began researching the effects of space travel and rocket launches on the body by launching rhesus monkeys on captured V-2 rockets. At the same time, the air force traveled to Africa where it captured sixty-five infant and young chimpanzees to further their research. These chimps were placed in dark, confined metal capsules and spun at high speed in centrifuges and strapped into rocket sleds where they were catapulted down a rail called the Daisy Track at more than three hundred miles per hour. At the end of the track, the brakes were slammed. These tests produced the results one would expect.

Chimpanzees were selected to certify the safety and performance of the capsules that would one day send an American astronaut into space because of their extreme intelligence, their ability to learn complex tasks, and because their physiological makeup is only one step away from that of a human. In late 1960, six chimps—four female and two male—were sent to Cape Canaveral and allowed to acclimate to the change in altitude and climate. Two of the chimps—Ham and Enos—were selected to test NASA's new Mercury capsule.

Ham—whose name was an acronym for Holloman Aero Medical—was said to be a jovial soul. His trainer recalls that he was fun loving, loved hugs, and was incredibly intelligent. Because of these traits, he was selected to be the first to test the combination of the new capsule and Wernher von Braun's Redstone rocket in order to certify them for human flight.

The three-year-old, thirty-seven-pound chimp was put through a rigorous training program—heat and pressure chambers, centrifuges, and isolation chambers. He sat in a custom fitted chair for five minutes a day for a week, the duration increasing with each subsequent week until he could sit on the couch for hours at a time.

A series of tasks were designed for the chimp to perform during the flight. A set of levers was mounted to a panel, each with a light above it. When a light came on, Ham would have to pull that lever. Pulling the incorrect lever

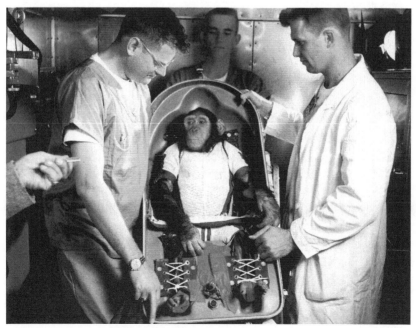

Three-year-old chimpanzee Ham in the biopack couch for his MR-2 suborbital test flight. *NASA*

resulted in a jolt of electricity delivered through metal plates attached to his feet. Pulling the correct lever, he was rewarded with either a snack or drink of water or fruit juice. In an interview, famed primatologist Jane Goodall said that she had talked with some of NASA's early astronauts about the shock training the monkeys went through, and they all told her that "they couldn't have gone through anything like that."

In the early morning hours of January 31, 1961, the Mercury capsule and Redstone rocket were readied. Several glitches in communications and tracking were corrected, and at 7:53 a.m., Ham, seated in his contoured couch inside his pressurized enclosure, was loaded aboard the capsule and sealed in.

Ham waited patiently in his cockpit as the launch was put on hold time and time again—an electronic inverter overheated, causing cabin temperatures outside of Ham's enclosure to rise to more than 180 degrees; the gantry elevator that provided access to the launch vehicle became stuck; and it took what seemed like forever to clear the pad of personnel.

Finally, at 11:55 a.m., Ham rocketed into the Florida sky. However, the thrust on the Redstone and the early release of the escape tower caused the rocket to fly to an altitude of 157 miles instead of the expected 115 miles, which resulted in the capsule landing farther downrange than planned and two hours away from the recovery ship.

Navy helicopters were dispatched to the scene where they recovered the bobbing capsule and flew it to the waiting recovery ship, *Donner*. Once onboard and removed from the capsule, Ham raced to jump into the arms of his trainer and happily accepted an apple and orange.

Upon returning to dry land, the media clamored around the astrochimp for photos. One of the most famous images shows Ham with what appears to be a huge grin on his face. Jane Goodall saw something different: "That was the most extreme fear that I have ever seen in a chimpanzee."[17]

Public concerns began to mount regarding the physical and mental condition of the chimp after his flight. NASA called a press conference in which it demonstrated that Ham would happily go back into his custom couch. It took four handlers to make the chimp lie down.

At this point, America should have been ready to become the first nation to send a human into space, but Redstone designer Wernher von Braun wasn't sure. After all, early Redstone test flights ended up becoming giant firework displays, with many exploding only seconds after liftoff, causing one of the future Mercury astronauts, Alan Shepard, to quip, "Well, I'm glad they got that out of the way."[18]

Von Braun ordered one more unmanned (and un-monkeyed) test flight. The flight was successful, but the Soviets surprised the world once again by launching cosmonaut Yuri Gagarin, who became not only the first human in space, but also the first to orbit Earth. The Americans had been beaten to the punch again. Five months after Ham's flight, Alan Shepard would take the chimp's same ride to become America's first man in space.

By November 1961, the United States was making plans to orbit its first astronaut, this time on the Atlas, another fully tested rocket that had a sketchy history. If all went well, the Atlas would be the rocket that would send John Glenn into Earth orbit and put America back in the space race.

On November 10, a one-and-a-half-pound squirrel monkey named Goliath was strapped into a capsule aboard the rocket at Cape Canaveral. A malfunction during the boost phase caused flight controllers to destroy the rocket thirty-five seconds into the launch, killing Goliath.

Only nineteen days later, another of the six Holloman chimpanzees would take a ride on an Atlas rocket. His name was Enos. According to his handlers, Enos was not a people person and would just as soon bite your finger off than show affection. His trainers couldn't carry him like Ham. He always walked on his own to training.

Enos was successfully launched into orbit on November 29. Like Ham, Enos was trained to respond to lights above a panel of levers. The chimp made one orbit of Earth in one hour and twenty-eight minutes, but there was a malfunction in the connection that produced the electric shock for incorrect lever responses. Enos performed his tasks flawlessly, but no matter what lever he pulled, he got an electric shock.

Enos splashed down successfully in the Atlantic Ocean and was quickly recovered. None the worse for wear, the chimp was returned to Holloman Air Base, where he died eleven months later from dysentery unrelated to his flight.

The United States and the Soviet Union were not the only two nations experimenting with animals in space. The French government launched the first cat in space from a launch site in the Sahara Desert. Felicette lifted off October 18, 1963, aboard a Veronique AGI sounding rocket, successfully reaching a suborbital altitude of one hundred miles on a ride that lasted fifteen minutes. Tragically, Felicette was euthanized eleven months later so scientists could "study her brain."

The final launch of a space monkey by the United States occurred just prior to the first moon landing in July 1969. Bonnie, a tiny pig-tailed macaque, was launched aboard the third of what NASA called the Biosatellite series.

During the flight, sensors attached to Bonnie showed that her heart rate and respiration were decreasing rapidly. She began to sleep more and for longer periods of time, and she refused to eat, losing 20 percent of her body weight. After only nine days of her thirty-day mission, the decision was made to fire the retro-rockets and bring her down early. Eleven hours after recovery, Bonnie died due to a heart attack.

Immediately following Bonnie's death, the nonprofit United Action for Animals published a lengthy editorial in its newsletter blasting one of the scientists in charge of the project, Dr. W. Ross Adey, for inhumane treatment of the monkey for implanting ten hairlike steel wires in Bonnie's brain to measure brainwaves. Adey filed a $2 million libel suit against the organization, which was operated by two women from their New York City apartment. The *New York Times* described the president of the organization, Eleanor Seiling, as a woman who lives on Social Security, "feeds stray cats in her upper West Side neighborhood and rolls her own cigarettes."[19]

The lawsuit was eventually dismissed.

By 1997, the chimpanzee research program at Holloman Air Force Base had 141 chimpanzees in its possession. Man had landed on the moon, and the chimps were no longer needed. The Air Force placed the colony up for bid. The winner was the Coulston Foundation, a "research" organization that had already been fined and sanctioned for mistreatment of animals for "the sake of science." Jane Goodall visited the facility and found the monkeys, kept in tiny cages, were exhibiting signs of mental illness—banging their heads against the walls and continually rocking back and forth.

After a lengthy battle, the nonprofit organization Center for Captive Chimpanzees (now Save the Chimps) won custody of twenty-one of these heroes, releasing them to roam free in a sanctuary in Florida. Since that time, a few others have been rescued but the majority were still housed at the Alamogordo Primate Facility. As of October 2020, thirty-seven chimps remain at the facility.

4

FROM THE MOON, INTERNATIONAL POLITICS SEEM PETTY

Orbiting above Earth every ninety-three minutes at an altitude of 254 miles, the International Space Station (ISS) is a gleaming example of what global cooperation looks like. Since the final module was installed in 2011, the station has hosted a total of 242+ people from nineteen different countries. The journey to get to this point has taken many twists and turns over the decades, with many countries vying for a place in space. Sometimes their efforts seemed futile compared to those of the space superpowers, while others made people shake their heads and wonder what they were thinking.

As early as 1964, other nations wanted to get into the space game, including the smallest of countries like the African nation of Zambia.

The nations of central Africa were once part of the British commonwealth, but at the start of the 1950s, dissatisfaction with British rule increased, spreading across the land and sweeping the population into various resistance movements. One of those rebels was Edward Makuka Nkoloso.

Following his service with the Northern Rhodesian Regiment in World War II, Nkoloso wrote a letter to the Rhodesian newspaper, the *Northern News*, clearly venting the feelings that were roiling in the hearts of Africans who had fought side by side with the British during the war. It was clear, as Nkoloso put it, that fighting for white men did not improve Africans' lives. "We are entirely forgotten."[1]

Nkoloso joined one of the Rhodesian resistance movements but was arrested in 1956 for his affiliation with it and imprisoned for one year. By the time of his release in 1957, Northern Rhodesia had effectively split from the commonwealth and become known as the Zambia African National Congress. Nkoloso became a politician and was elected as the security official for the newly formed United National Independence Party.[2]

Eventually, the fledgling nation completely broke ties with the commonwealth, gaining complete independence in 1964 as the Republic of Zambia. Being a newborn country trying to find its legs in the world, Zambia's sudden announcement to the world that it was forming its own space program was an incredible leap.

Well, that's not quite true. The so-called space program was not government sanctioned. It was conceived by none other than Edward Nkoloso. In 1960, Nkoloso founded the Zambia National Academy of Science, Space Research, and Philosophy and began lobbying for a Zambian space program with the goal of landing a group of "Afronauts" on the moon and, eventually, twelve more Afronauts and ten cats on Mars.

Days after Zambia secured complete independence from Britain and while the joyous celebrations of freedom swept across the new nation, Nkoloso lamented in a local newspaper that the celebrations were interfering with his quest for space:

> The rocket could have been launched from Independence Stadium and Zambia would have conquered Mars only a few days after independence. Yes, that's where we plan to go—Mars. We have been studying the planet through telescopes at our headquarters and are now certain Mars is populated by primitive natives. Our rocket crew is ready. Specially trained space girl Matha Mwamba and two cats (also specially trained) and a missionary will be launched in our first rocket.[3]

Time magazine was the first publication to bring Nkoloso's story to the world in an article that focused on the fledgling country's new president, Kenneth David Kaunda. The article made only a brief reference to Nkoloso's dream of a Zambian space program and the Afronauts' would-be training.[4] That single paragraph about Nkoloso spurred interest among other news outlets around the world, including the Associated Press, which interviewed Nkoloso. "Some people think I'm crazy," Nkoloso said. "But they'll be laughing when I plant Zambia's flag on the moon."[5]

His Afronauts underwent rigorous training in the most unconventional of ways: they were placed in empty fifty-five-gallon oil drums and rolled downhill to simulate training in a centrifuge; trainees would swing out far and high on a rope swing and, upon reaching the highest point of the arc, the rope would be cut, giving momentary—very momentary—weightlessness; and they walked on their hands because Nkoloso said that was the only way a human could walk on the moon. The rocket itself was a copper and aluminum tube that would be launched via catapult.

Needless to say, the Zambian space program never got off the ground. Nkoloso requested more than $2 billion from UNESCO to fund the program. His request was never answered. Not long after, the trainees began demanding to be paid, and the lone female trainee, sixteen-year-old Matha Mwamba, became pregnant and her parents ushered her home.

Some people say that Nkoloso was "Zambia's village idiot,"[6] whereas others believed his training base was actually a safe house for freedom fighters who were still rebelling against British rule in other countries. Either way, the Zambian space program was a unique entry in the history of space exploration. And although Zambia never created a space program, fourteen other African nations did, beginning with the Egypt Space Agency, which launched the continent's first telecommunications satellite—NileSat 1—in 1998. Since that time, the other thirteen nations have either launched satellites or provided research for other African space agencies, spending more than $3 billion in the process, all in the name of space exploration and satellite technology.[7]

In 1959, one year after the National Aeronautics and Space Agency was founded, the agency invited other nations of the world to join it in exploring space during a meeting of the Committee on Space Research (COSPAR). One of the first countries to sign on was Great Britain. The British Science and Engineering Research Council (SERC), with assistance from NASA's Goddard Spaceflight Center, developed the Ariel 1 satellite.

The twenty-three-by-nine-inch cylinder was launched on April 26, 1962. During its lifetime, Ariel experienced several technical glitches that caused data to be transmitted back to Earth sporadically, but overall, it operated for two years measuring the effects of the sun's radiation on the ionosphere.[8]

From its very inception, NASA was not averse to international cooperation when it came to the exploration of space. President John F. Kennedy took that effort one step further, not only proposing an accelerated program with American allies, but also a collaboration with the Soviet Union, going as far as to propose a joint effort to land men on the moon.

Following one of the closest elections in US history in 1960, President Kennedy prepared to make his first State of the Union address to a joint session of Congress and the world. Tucked away in the middle of that soaring speech rallying Americans to a better future while warning of the perils and difficulties that lay ahead, Kennedy extended the first invitation for a joint space venture with the Russians:

> Today, this country is ahead in the science and technology of space, while the Soviet Union is ahead in the capacity to lift large vehicles into orbit. Both nations would help themselves as well as other nations by removing these endeavors from the bitter and wasteful competition of the Cold War. The United States would be willing to join with the Soviet Union . . . in a greater effort to make the fruits of this new knowledge available to all.[9]

Keep in mind that neither country had launched a man into space at this point—only unmanned scientific and communication satellites—but of course, that all changed on April 12, 1961, when cosmonaut Yuri Gagarin became the first person to orbit Earth. Only one month later and under extreme pressure from the world to have the United States catch up and overtake the Soviet Union in space,[10] Kennedy set the nation on a path to the moon when he announced during a special joint session of Congress that the nation should achieve the goal of "landing a man on the moon and returning him safely to Earth."

Nothing much came of Kennedy's first invitation to the Russians, mainly due to the fact that the United States lagged much farther behind the Soviets in the space race. After Gagarin orbited Earth, the United States entered the race by launching two astronauts—Alan Shepard and Gus Grissom—on fifteen-minute suborbital flights. Albeit groundbreaking for the Americans, it was less than inspiring to the Russians, giving them even more room to gloat about Gagarin's historic flight. It would take the United States ten months after Gagarin to finally launch John Glenn into orbit on an Atlas rocket. Once Glenn had set his boots on the recovery ship marking the end of his own history-making flight, Prime Minister

Khrushchev sent a congratulatory letter to Kennedy. In it, he finally responded to the president's original outreach:

> If our countries pooled their efforts—scientific, technical, and material—to master the universe, this would be very beneficial for the advance of science and would be joyfully acclaimed by all peoples who would like to see scientific achievements benefit man and not be used for "cold war" purposes and the arms race.[11]

Kennedy found Khrushchev's response encouraging and tasked NASA with writing a list of concrete proposals that both countries might be able to agree upon and eventually make a reality. The White House and State Department massaged that list into a more diplomatic response and sent it off to the Kremlin. In the proposal, the United States suggested that the two countries establish a joint global weather satellite network, exchange spacecraft tracking technology and data, assist each other and the world in mapping Earth's magnetic field, and jointly develop international telecommunications satellites. He further suggested that they "might cooperate in unmanned exploration of the lunar surface, or we might commence now the mutual definition of steps to be taken in sequence for an exhaustive scientific investigation of the planets Mars or Venus, including consideration of the possible utility of manned flight in such programs."[12]

Soviet reaction was positive and for the next year, NASA Deputy Administrator Dr. Hugh Dryden, the director of the agency's Office of International Programs Dr. Arnold Frutkin, and Soviet diplomat and space scientist Anatoly Blagonravov hammered out an agreement that basically followed President Kennedy's proposal to develop weather and scientific satellites.[13] A speed bump in history, however, brought progress on any collaboration in space to a grinding halt when it was discovered that Soviet nuclear missiles were being deployed just off the coast of Florida in Cuba. For the next thirteen days in October 1962, the world was gripped in fear that there would be all-out nuclear war.

The Kennedy administration decided to delay announcing the agreement, saying that there would be no further action on the US-USSR outer space collaboration until the crisis was over. When the missiles were finally removed from the island, the world breathed a sigh of relief while still reeling with the realization of how close it had come to nuclear annihilation. The specter of that threat hung heavy over the world forevermore, making the Cold War even chillier.

The announcement of the space agreement was delayed for two months until it was finally made public on December 5, 1962, at the United Nations, but now, the effort had taken on a dark and extremely political feel. While Dryden was still brokering for an honest and fair joint space program, the Russians began to drag their feet, interested only in programs that would directly benefit themselves.

The following year, Kennedy attempted to put the train back on the tracks by making a new overture in an attempt to nudge the Russians into joining forces. It was in an address he made to the eighteenth United Nations General Assembly on September 20, 1963:

> Space offers no problems of sovereignty, by resolution of this Assembly, the members of the United Nations have forsworn any claim to territorial rights in outer space or on celestial bodies, and declared that international law and the United Nations Charter will apply. Why, therefore, should man's first flight to the moon be a matter of national competition? Why should the United States and the Soviet Union, in preparing for such expeditions, become involved in immense duplications of research, construction, and expenditure? Surely, we should explore whether the scientists and astronauts of our two countries—indeed of all the world—cannot work together in the conquest of space, sending someday in this decade to the moon not the representatives of a single nation, but the representatives of all of our countries.[14]

The ceiling of the Payload Operations Integration Station at the Marshall Spaceflight Center in Huntsville, Alabama, is aglow with the flags of the many nations that are making the International Space Station a success. *NASA/MSFC/Emmett Given*

In the end, other than an exchange of a small amount of weather and magnetic measurement data, nothing of significance came of Kennedy's efforts. Two months after his UN speech, President Kennedy was assassinated and never witnessed the cooperation that would eventually unfold. It would take another decade before the two nations opened a hatch and shook hands in space during the Apollo-Soyuz Test Project that followed the Apollo program. Years later, the countries performed joint operations aboard the space stations Mir and the International Space Station.

During a congressional hearing in 1965, Dryden summed up the attitudes of both countries in the early 1960s and their inability to make joint operations a reality:

> I would describe the situation as a form of limited coordination of programs and exchange of information rather than true cooperation. They have not responded to any proposals which would involve an intimate association and exposure of their hardware to our view.

One congressman asked Dryden if the prospect for future cooperation was based on continued competitiveness and politics. He replied, "As near as we can tell at the moment."[15]

As the 1960s rolled on and the Americans and Russians dropped all notions that the two could be joint partners in space—at least for the moment—international cooperation among other nations was just beginning. Following the launch of the collaborative Ariel satellite on April 26, 1962, by the United States and Great Britain, the door to space was thrown wide open, and other nations followed in quick succession. This new international space community was led by Canada, which became the third country to design and build its own satellite, Alouette 1. Launched on September 29, 1962, from the Pacific Missile Range in California aboard a Thor-Agena rocket, the mission was a tremendous success. The rocket placed the satellite into a near-perfect circular orbit, and though scientists predicted it would have a life expectancy of only one year before deorbiting and burning up in Earth's atmosphere, the tenacious little electronic package surprised everyone and kept reporting back to ground stations for ten years.[16]

During the ensuing years, Japan and China joined the exclusive space club followed by others. It had become clear that there would need to be a set of rules and regulations to govern and protect this new frontier. The United Nations already had been working to this end and laid out the groundwork for a treaty during a meeting of the General Assembly in 1963. The resulting document was called the Outer Space Treaty.[17] It was eventually signed by the United

Listeners of Canada's international shortwave station, Radio Canada International, received a special card commemorating the country's satellite history. *Joe Cuhaj*

States, Soviet Union, and United Kingdom in January 1967 and put into effect in October of that year.

The basic tenets of the document stated:

- Space exploration would be for all mankind.
- Outer space should be free of exploitation by any country.
- The moon and planets should be explored peacefully.
- Nuclear weapons or weapons of mass destruction should not be placed on, in orbit of, or used on a celestial body.

That last bullet point in the treaty banning the use of nuclear weapons in space was an obvious and common-sense inclusion but may have been added in part as a response to a discussion by Russian scientists in 1958 concerning their plans to land a man on the moon. The discussion centered around a possible outcome of successfully landing a human on the moon, an outcome that the American space program never considered but perhaps should have—that some of the public would think the moon landing was a hoax.

Immediately after the successful launch of the world's first artificial satellite by the Soviet Union in 1957, the Russians set their sights on loftier and ever more daring goals in space. The ultimate demonstration of their technological prowess would be a manned landing on the moon.

Only two years after launching Sputnik 1 and orbiting the dog Laika, the Soviets sent the first probes beyond Earth's gravitational pull—Luna 1 and Luna 2. In addition to studying interplanetary material and gases, cosmic rays, and meteorite particles during the voyage, the spacecraft would be the first to either fly by or orbit the moon. Because the tracking systems of the day were rudimentary at best, the spacecraft would release a cloud of glowing orange sodium gas[18] that could be spotted by astronomers on Earth, thus showing the spacecraft's position.

Luna 1 was launched on January 2, 1959, and traveled within 3,700 miles of the lunar surface before settling into an orbit around the sun somewhere between the orbits of the Earth and Mars. Luna 2 became the first man-made spacecraft to land on another world, albeit a crash landing on the lunar surface. Despite the crash, the Russians had proven that their technology was rock solid and that they could proceed with their lunar ambitions. But in the back of their minds, Soviet scientists had a nagging feeling about a lunar landing: Once you land, how can you prove to the world that you actually did it? No one would be able to see it on Earth and the spectacular feat would be labeled a hoax (sound familiar?).

With hindsight, that is a fair and intriguing question. After all, here we are, fifty-plus years after Neil Armstrong and Buzz Aldrin became the first humans to set foot on the moon and despite television images (albeit grainy black and white) of the monumental event being beamed live into just about every living room across the globe, polls show that 6 percent of Americans[19] believe the landing was a hoax. These nonbelievers tend to be vocal as Aldrin discovered in 2002 when he was confronted by one who got into the astronaut's "space" and called the second man on the moon a liar. The former astronaut had had enough and punched the man in the face.[20]

So, if the Russians could land a person on the moon, how could they prove it to the world? Their answer? Nuke it.

In 1958, the US and Russia already had become world superpowers by perfecting the atom and hydrogen bombs. One of Russia's top rocket scientists, Boris Chertok, suggested that they should detonate an atomic bomb near the lunar landing site to prove that they had made it.

In a 1999 interview with the London newspaper the *Independent*, Chertok said, "The plan was to send an atomic bomb to the moon so that astronomers across the world could photograph its explosion on film."[21]

After a series of long, deliberate, and serious discussions on the subject, the idea was tossed into the dustbin of history. Scientists realized that since the moon had no atmosphere, there would be no telltale mushroom cloud or hellish red fireball. "The flash would be so short lived," Chertok said, "that it might not even register on film."

And what about the moon? What would happen to it if a nuclear explosion occurred? Unlike the 1970s sci-fi television show *Space 1999*, in which the moon is blasted out of its orbit due to a nuclear accident, a real-life bomb would leave only a small crater.

Since those early formative days and many attempts to bring the world together for the common good through space exploration, international cooperation today is routine. Since the Outer Space Treaty was enacted, countless other cooperative agreements have been signed between nations. As of 2020, more than 650 people from forty-one countries have soared into the final frontier aboard either a US space shuttle or a Russian Soyuz spacecraft to the Russian Mir space station and later to the International Space Station. What is interesting about that fact is what each country calls their space travelers.

Most of the individuals who have flown into space have been labeled with a moniker that ends with the suffix *-naut*, which is derived from Greek and translates to "sailor." With the prefix *astro-* (meaning "stars"), the word *astronaut* translates to "sailor to the stars."

The first appearance of the word *astronaut* is believed to have been in the 1880 book, *Across the Zodiac*[22] by British author Percy Greg. The word was used as the name of a fictional spaceship and was a tip of the hat to the Argonauts ("sailors of the Argo") found in the Greek mythology of Jason and the Argonauts.

As the United States forged ahead with its manned space program, it needed to give a name to its space travelers. NASA was in a bit of a quandary as to what it should call its space men. According to one of the agency's first directors of manpower, Allen O. Gamble, space travelers originally were going to be called "Mercury," a nod to the Roman messenger to the gods. You guessed it: that name had been spoken for already by the first American space program. As Gamble explained, "With our best name already taken, out came the dictionaries and thesauruses. Someone found that the term 'aeronaut,' referring to those who ride in balloons . . . was derived from 'sailor in the air.' From this we arrived at 'astronaut,' meaning, 'sailor among the stars.'"[23]

The Russians had a similar problem. They began thinking about what to call their pilots as soon as Sputnik roared off the launch pad in 1957. They toyed with using "astronaut" then "pilot cosmonaut." A short three-paragraph article buried in the *New York Times* on April 13, 1961, reported that it was Prime Minister Khrushchev who made the final decision—they would be called[24] "cosmonauts," or "sailor of the universe."

When the Chinese National Space Administration launched its first pilot, Yang Liwei, into space on October 15, 2003, he was called a *hang yuan* by his countrymen. Westerners called him a *taikonaut*, the prefix deriving from the Chinese word *tai kong*, which means "great emptiness."

Then we add the French to the mix, who call their astronauts *astronaute* or sometimes *spationautes*.

The bottom line is that they are all the same: astronauts, cosmonauts, *taikonauts*, *spationautes*. All the same. The takeaway from all of this is that no matter what you call them, they are all global heroes.

5

$12 A DAY TO FEED AN ASTRONAUT: WE CAN FEED A STARVING CHILD FOR $8

It wasn't a particularly cold day as far as winters go in New York City. On the Sunday before Christmas 1965, kids in the New York City tristate area awoke to their normal Sunday morning routine, grabbing a quick bowl of cereal, oatmeal, or Cream of Wheat and plopping down in front of a behemoth of a console television to watch the grainy black-and-white images of their favorite television show, *Wonderama*.

The television show was standard kids' programming of the day, airing a good smattering of Looney Tunes cartoons, barraging the kids with commercials for the latest toys, maybe offering prizes that kids could win, but there was a twist—its host.

Sonny Fox was a tall, lean man who hosted the show from 1959 to 1969. His demeanor wasn't goofy and silly, but he captivated children because he talked to them as if they were adults. He never talked down to them and always had fascinating and inciting interviews with artisans, entertainers, and politicians. He was called the "Johnny Carson of the elementary years."

On this Sunday morning, Senator Robert F. Kennedy was the show's special guest. Kennedy had appeared on the show before, and each time, the children in the live audience were quiet and well-behaved as they listened intently to what the senator had to say.

During this interview, a bespectacled young man leaned in from his seat on the floor and asked an incredibly intelligent and provocative question, one that many adults, particularly in the black community, were asking:

"Do you think that all the money we've been spending on this nation's space program should be spent on this or poverty bills?"

Kennedy didn't hesitate with his answer, "I think we can make the space effort. It's worthwhile. It's the exploration of the atmosphere. If there is ever an unknown, man is going to search the unknown, so I think that's worthwhile and I think we can [do] the other programs as well."[1]

That brief encounter between Kennedy and a young schoolboy illustrates what is often overlooked by the media and textbooks: the public was not quite as enamored with the space race as the media portrays. In fact, a 1961 Gallup poll showed that 58 percent of Americans did not want to spend $40 billion to send a man to the moon and instead would rather spend the money on social issues facing us on Earth.[2] Those polling numbers didn't fluctuate much all the way through the moon landing in 1969.

Additionally, the media made it as if the US space program was a self-contained bubble, far removed from events that were taking place back home on Earth in the 1950s and 1960s—the fight for desegregation, voting rights, the war on poverty. However, nothing could be further from the truth. Even before NASA became a reality, its predecessor—the National Advisory Committee for Aeronautics (NACA)—felt the winds of social change. Within its workforce, a peaceful "protest" was underway. There were no marches. There were no sit-ins. It was a quiet movement that the public was unaware of. Despite some resistance and headwinds from within—even from the federal government itself—the agency began to integrate its ground operations and facilities, albeit very slowly.

The first attempts at squelching discriminatory practices in the federal government came prior to World War II. In June 1941, President Roosevelt signed Executive Order 8802, which banned discriminatory practices by any federal agency, union, or company engaged in war-related work. Two years later, at the height of the war, NACA and its researchers at the Langley Memorial Aeronautical Laboratory in Hampton, Virginia, found themselves backlogged with an almost crippling workload. Data from their aeronautical research was becoming unmanageable, overwhelming the staff to the point where the agency made the decision to open its doors and begin desegregating the workforce, even on a minor scale, by hiring black mathematicians to help process the load.

In 1943, a young black math teacher from Robert Russa Moton High School in Farmville, Virginia, was hired and assigned to the research center's segregated West Area Computing Unit. Her name was Dorothy Vaughn. She was to be what NACA called a "computer"—a person who reads, calculates, and plots the data being accumulated at Langley during its aerodynamic research on aircraft and missiles in wind tunnels.

Vaughn and her fellow computers did an extraordinary job under extraordinary conditions, quietly fighting Jim Crow laws prohibiting the team from dining with their white counterparts and using the same bathroom facilities as well as facing racist attitudes from many of their counterparts.

Despite these obstacles and without complaint, Vaughn and later Katherine Johnson (who joined NACA in 1953) and Mary Jackson (who was hired in 1958) played integral roles in helping humans reach space (and all three were the focus of the best-selling book and hit movie *Hidden Figures*). Vaughn became the lead of the West Area Computing Unit in 1949, assisting in the compilation of a handbook for algebraic methods used in calculating machines, and an expert in one of the early computer languages, FORTRAN. Throughout her career, which ended with her retirement in 1971, Vaughn was a staunch advocate for the rights and promotion of not only black women but also white women at NACA and later NASA.

Johnson's accomplishments included calculating the trajectory for America's first flight into space with Alan Shepard and its first orbital mission with John Glenn. For Glenn's flight, NASA had just installed new state-of-the-art electronic computers to perform the calculations. Unsure of the accuracy of the new machines, Glenn demanded that NASA, "Get the girl [to calculate it]. If she says they're good, then I'm ready."[3]

Johnson also calculated the trajectory for spacecraft to send it on its way to the moon and the first manned lunar landing.

Mary Jackson worked at Langley as a computer as well and eventually was promoted to work in the center's wind tunnel research program. With hands-on research under her belt, her supervisor suggested she take courses at a local school to meet the necessary requirements to become an engineer. After a tussle with the local school board, Jackson was permitted to enroll in the all-white school, completing all required courses to become NASA's first black female engineer.[4]

Similar stories, like that of Julius Montgomery's arrival at Cape Canaveral, demonstrated the barriers that slowly were being nudged away at other NACA facilities.

Montgomery was the oldest of Edward and Queen Ester Jackson Montgomery's twelve children. The Homewood, Alabama, native suffered from severe stuttering, but with determination, he overcame the disability, which gave him the confidence to enroll at the Tuskegee Institute, where he graduated in 1951 with a bachelor's degree.

After college, Montgomery joined the Air Force, where he earned a first-class radio telescope operator's license. After his discharge, he was hired as a "range rat" at the RCA development laboratory at Cape Canaveral, Florida, where he repaired radar, missile, and satellite electronics. In their book, *We Could Not Fail: The First African Americans in the Space Program*, authors Richard Paul and Steven Moss summed up Montgomery's first day on the job succinctly, calling it "a dispiriting combination of the sad and hateful present, tinged with the bitter history of yesterday."[5] As they put it, he was the first African American hired as anything other than a janitor at the Cape.[6]

All alone with no backup, no major national movement urging him on, no press to cover the moment, Montgomery walked into the building where he met his new coworkers, all of whom were white. As he passed his fellow engineers, they turned away and refused to look at him. "Nobody would shake my hand," he recalls. "I got to the last fellow, and I said, 'Hello, I'm Julius Montgomery.' He said, 'Look boy, that's no way to talk to a white man.' I said, 'Ah, forgive me, oh great white bastard. What should I call you?' I laughed, he laughed, and he shook my hand."[7]

Montgomery later went on to become the first black graduate of the Brevard Engineering College, which is now the Florida Institute of Technology.

The fight for civil rights is nothing new in the United States. Since its inception, the country has been battling to keep the promises made by its founders that all men are created equal. But "civil rights" is an umbrella term with many offshoots, all equally important, including poverty, homelessness, unequal educational opportunities, voting rights. Civil rights is an undercurrent that flows through everyday life in the United States, and many people are ambivalent to its corrosive and consequential effect on society and the human toll it takes. In the 1950s and 1960s, the voices of those trying to promote social change began to rise and be heard.

During this time of tumult, countries around the world watched the American experiment in democracy work its way through these growing pains and wondered if it would survive. One of America's adversaries used such moments of unrest against the United States as propaganda demonstrating that the promises of America were lies and that democracy was failing its people. That is where the junction between the exploration of space and the social upheaval and unrest of the 1950s and 1960s first intersected.

Across the American South of the 1950s, Jim Crow laws made segregation of the public school system mandatory. African American children could not attend the same schools or obtain the same level of education afforded white children, which placed them at a stark disadvantage from an early age. On May 17, 1954, the US Supreme Court ruled in the landmark case Brown vs. Board of Education of Topeka that segregation of the country's public schools was unconstitutional. The response by southern states was immediate, forceful, and defiant. The ruling only strengthened the white South's resolve and its segregationist ways; it would not stand for this government overreach.

The court also issued a second ruling known as Brown II, which ordered schools to integrate with "all deliberate speed." With pressure from the National Association for the Advancement of Colored People (NAACP), the Little Rock, Arkansas, school board agreed to a gradual integration of its schools. Nine African American students registered for classes in September 1957. The governor of the state, Orval Faubus, put the federal government on notice that he would deploy the Arkansas National Guard for the "safety of the students," insisting that there would be tremendous violence once the nine students set foot on the campus of Little Rock's Central High School.

The children's parents agreed that all nine students would carpool to the school together and enter the building as a group, but one of the students, fifteen-year-old Elizabeth Eckford, did not have a telephone and did not receive word of the plan. On September 4, 1957, Elizabeth arrived at the school alone, a notebook clutched to her chest as she approached the school among a gathering crowd of white students and townspeople who spat on her and hurled verbal abuse.

Instead of helping the students into the school, the National Guard blocked the nine from entering. President Eisenhower ordered troops from nearby Fort Campbell to take control of the guard, and the students were eventually allowed to enter. For the remainder of the school year, the students faced verbal abuse, the burning of black effigies, and physical violence despite a military presence.

As is often the case, the story was a flash in the pan for the news media, a blip on the screen, and just as quickly as it had appeared on television sets across the country, the media lost interest and the story disappeared. Eisenhower's administration, however, knew that the story was not over as they watched it reverberate around the world. Secretary of State John Foster Dulles was quoted in the *Arkansas Gazette* as saying the story of the Little Rock Nine was "not helpful to the influence of the United States abroad."[8]

Dulles was right. As the story went cold for journalists in the United States, it was still burning hot around the world, especially in the Soviet Union, which used the event as propaganda showing that democracy was a failed experiment. Headlines in Soviet newspapers mocked the hypocrisy of the United States, preaching freedom for all while denying it to many. The communist youth newspaper *Komomsomolskaya Pravada* blazed the headline "Troops Advance against Children!"[9]

The Soviets were afforded the opportunity to rub salt in America's wound when they launched Earth's first satellite, Sputnik 1, on October 10, 1957. In announcing to the world the sphere's successful flight, the Soviet news agency, TASS, boasted that their socialist form of government was far superior to capitalism, stating, "Artificial Earth satellites will pave the way to interplanetary travel and, apparently our contemporaries will witness how the freed and conscientious labor of the people of the new socialist society makes the most daring dreams of mankind a reality."[10]

The worldwide voice of the Soviet Union, Radio Moscow, went a step further and linked Sputnik and the forthcoming space race to the events in Little Rock, tweaking America's nose by announcing the times that the space ball would fly over the Arkansas city. This forcefully contrasted the technological prowess of the Russians over the United States, which they said was "locked in racial strife as armed federal soldiers sought to defend nine black students from a baying white mob." The announcers admonished Americans, telling them to think of civil rights as a foreign policy and not just a domestic policy.[11]

Americans were not concerned about what the Russians thought of our form of government. All the public could see was that the enemy had scored a big technological victory with Sputnik and a huge military advantage over the United States. To add insult to injury, only a few weeks later, the Soviets launched Sputnik 2 with the dog Laika onboard.

The feeling on Capitol Hill was that the Soviet space feats posed an imminent military threat to the United States and the free world, completely ignoring the fact that these were incredible feats of science. Senate Majority Leader Lyndon B. Johnson was the most vocal, comparing the dual Sputniks to Pearl Harbor. Johnson—who was the chair of the Preparedness Investigating Subcommittee of the Senate Armed Services Committee—scheduled hearings on the military implications of the Sputniks. The hearings convened on November 25, 1957, and after five weeks of testimony from seventy-three scientists and engineers, resulting in a 1,300-page document, Johnson's views on spaceflight drastically changed. According to one of Johnson's aides and a Legal Reference Service analyst, Eilene Galloway, "[The hearings] changed the perception of the problem from one that was originally only national defense to one that also had beneficial uses from space and meant that we could hope for peace."[12]

Johnson became a staunch advocate for the US space program, pushing forward a proposal that would create NASA one year later in 1958. Johnson later was tapped to be John F. Kennedy's vice-presidential running mate, and after they had won the 1960 presidential election, Kennedy assigned Johnson to head the National Aeronautics and Space Council and the administration's Committee on Equal Opportunity.

If there was one thing Kennedy and Johnson agreed upon, it was that the root of racial injustice in the South was brought on by southern poverty. The only way to stem the tide was to afford minorities economic opportunities that never existed before in the South—highly technical, cutting-edge, and higher-paying jobs. Only six weeks after the first human—Yuri Gagarin—orbited Earth and two weeks after Alan Shepard took his historic ride into space on *Freedom 7*, Kennedy made a bold request of Congress: to send an American to the moon by the end of the decade. This proclamation and subsequent approval by Congress opened the door for Johnson, who saw the space program as the key to ending southern poverty, or at least a stepping-stone to that end.

Johnson set about establishing NASA centers to facilitate the moon landing across the South in Alabama, Florida, Mississippi, Louisiana, and

Texas. The Apollo program and the projects that preceded the lunar missions—Mercury and Gemini—would require an enormous work force of more than two hundred thousand. Suddenly, these new NASA facilities found themselves with an incredible number of job openings to fill, and in 1963, the Marshall Spaceflight Center (MSC) in Huntsville, Alabama, began to search for young black engineers to fill those positions.

To assist in the effort, MSC hired Charlie Smoot, who quickly established a new cooperative education program in which black students alter-

Charlie Smoot created a co-op training program for young black engineers at the Marshall Spaceflight Center in Huntsville. *NASA/MSFC*

nated semesters between working on their degree in the classroom and then interning at Marshall the next. Smoot began by paying visits to the country's six black colleges that offered accredited engineering degrees from which to recruit this new generation of engineers.

Smoot, who has been called NASA's "first Negro recruiter" by historians, garnered the aid of Marshall's first director, Dr. Wernher von Braun, to aid in the search, dispatching the rocket scientist to colleges where he proclaimed the benefits of working at Marshall and the wealth of practical experience the students would gain by signing on.

In an interesting side note, von Braun, who had been chastised by some for allowing the use of slave labor to build his V-2 rocket for the Nazis in World War II, was staunchly against discrimination, likening laws like poll taxes, which prevented many blacks from voting, to putting the Berlin Wall around the ballot box. Von Braun declared, "I am not going to sit quietly on a major issue like segregation."[13]

While the opportunity to train on the cutting-edge technology that would send a man to the moon was an extraordinary opportunity, life outside MSC was a challenge for the black students who arrived in Huntsville. In a 1963 document produced by NASA, the agency required that "installation offices placing reservations at hotels, motels, restaurants and recreational facilities shall limit such reservations to those places of public accommodation where NASA employees, other Government employees and guests of NASA will be served without discrimination as to race, color, creed, or national origin."[14]

On paper, that was all well and good, but putting that edict into practice was another story. The fact of the matter was that hotels were segregated across the South, especially in Alabama. NASA had to find local black families to board the students during their stay.

Two African American students—an electronics engineer working at the Redstone Arsenal and a mathematician at MSC—were denied admission into the University of Alabama Huntsville Center because of a "technicality." The *New York Times* reported that university officials were unwilling to "admit Negroes without a further court order because of the defiant position taken by Governor George C. Wallace. He has vowed repeatedly he would personally thwart any step toward desegregation."[15]

On January 20, 1962, four young African American students from Alabama A&M—William Pearson, Leon Felder, Bertha Burl, and Mary

Joyner—along with NASA technical artist Marshall Keith (who was white), strolled into a Huntsville Woolworths where Keith ordered an egg breakfast. When the waitress served him, he shoved the plate down the counter where it came to a stop in front of one of the black students, Leon Felder, who began eating.

The restaurant manager was enraged that the black students had been served. He walked up to Felder, who was enjoying the breakfast and asked, "Want more salt and pepper?" and then proceeded to pour the entire bottle of the condiments on the meal. Felder kept eating.

The manager then dumped an entire bottle of ketchup on the counter in front of the girls. They simply picked up napkins and began cleaning up the mess. The manager went into a rage and slammed plates on the floor.

A few days later, Keith was relaxing at home when there came a knock on his door. When he answered it, he was greeted by two armed, masked men who blindfolded the artist, marched him to their car, and drove him to a remote location where they stripped him naked and doused him with pepper spray, causing severe burns to his skin. Keith managed to make his way to a nearby house owned by an African American family who rushed the man to the hospital for treatment.[16]

As the newborn space agency grappled with segregationist laws that impeded the growth of their centers across the South, externally, civil rights leaders continued their fight for voting rights, integration, and the fight against poverty. Despite not being overwhelmingly in favor of the space program, after Yuri Gagarin became the first person to orbit Earth in 1961, Americans—for the moment, at least—were more concerned about being upstaged by the Russians than the plight of black Americans on the ground. In his memoir, *Walking with the Wind: A Memoir of the Movement*, the late civil rights leader and former congressman John Lewis remembered the month of May 1961 when Alan Shepard became the first American in space. As Shepard made his fifteen-minute suborbital flight, scores of black and white volunteers hopped aboard Greyhound buses and fanned out across the South to test the Supreme Court's ruling that made segregated facilities for interstate travelers illegal.

Those Freedom Riders headed out across the South, and on Mother's Day, one of the buses was met in Anniston, Alabama, by an angry white mob that viciously attacked the riders. They slashed the bus's tires, smashed its windows, and lobbed a firebomb through a window, setting it on fire and causing the gas tank to explode. Shots fired by state police drove the mob back, giving the riders enough time to escape with their lives. The same fate met other Freedom Riders across the South, including Lewis and those who rode into Rock Hill, South Carolina.

Lewis wrote in his memoir about his concern that Shepard's flight and resulting celebrity upstaged the message of the Freedom Riders.

> The next morning's newspapers across the country carried a small story about the beating of the Freedom Riders in South Carolina. There's no telling, though, how many Americans paid attention to that little story, considering that the day's big headline was the flight of NASA's first manned rocket. . . . I had no idea that history was being made up in space while I was getting beaten in Rock Hill.[17]

At that moment in time, Lewis was correct. These two monumental moments in American history were running a parallel course, with the American space program seemingly upstaging and obscuring the message of the other at every turn. But in 1965, television finally became a powerful and transformative force, bringing the message of these protestors into sharp contrast with the technological marvels of spaceflight.

It happened on March 7, 1965, when more than six hundred protesters convened in Selma, Alabama, to begin a march from the sleepy central Alabama town to the state capital some fifty miles away to demand their right to vote in the state. Television cameras transfixed the nation, transmitting the harsh reality of what was happening in the South and the brutality and inhumanity perpetrated on the marchers.

As John Lewis led the procession and the marchers were about to leave town crossing the now iconic arched steel Edmund Pettus Bridge, they were met by state troopers and local law enforcement who beat them unmercifully with billy clubs and choked them with tear gas, driving the marchers back across the river into town. The event, which became known as Bloody Sunday, was broadcast into homes around the country and across the world, outraging the nation and galvanizing the civil rights movement in America.

A few weeks later, the United States was ramping up its race to the moon as it prepared for the first flight of the two-man Gemini program. Onboard the cramped Gemini 3 capsule was one of the original Mercury 7 astronauts,

Gus Grissom, and one of the "new twelve," John Young. Television cameras were there to cover the event. Journalist Jack Gould of the *New York Times* was quick to pick up on the moment, reminding readers that television coverage showed "two astronauts who orbited the earth three times while marchers covered a few miles. . . . On television, it was a day when a viewer had no doubts of the contrasts of 1965."[18]

Following the events of Bloody Sunday, Alabama Governor George C. Wallace feared for the state's economy and wanted to present the state as being much more than "segregation now, segregation tomorrow, segregation forever" as he had proclaimed during his inauguration in 1963. To do so, Wallace organized a junket for the press and businessmen in June 1965 to tout the state. The See the Truth tour took Wallace's guests to cities and towns across the state with a final stop in Huntsville and the Marshall Spaceflight Center.

The center's director began preparing for the visitors and did what he did best—public relations. Von Braun arranged for a test firing of one of the enormous Saturn V rocket engines that would eventually send a man to the moon. The invited guests stood in awe and watched as the powerful engine burst to life with a hellish red flame that devoured the test stand's fire deflector. A few short minutes later, the engine shut down, leaving nothing but an enormous cloud of smoke.

The onlookers were instructed to wait in the viewing area until it was deemed safe for them to leave. Von Braun saw his chance and made a speech to his captive audience while giving Wallace a gentle slap in the face:

> If we are to keep our best people from leaving the state and [encourage them] to become the leaders of tomorrow, it is mandatory that we be able to offer them the same opportunities here that they can find in any other state. The promise of the era will not be recognized by those who stand and wait. The era belongs to those who can shed the shackles of the past.[19]

Wallace never returned to Huntsville for the duration of the Apollo program.

Following the passing of the Civil Rights Act of 1964, prohibiting discrimination of any kind, including race, color, religion, sex, or national origin, and the Voting Rights Act of 1965, civil rights leaders cast their attention firmly on the issue of poverty in America. By 1966, the movement had joined the voices of the 58 percent of Americans who thought that money spent on the space program could be better spent here on Earth. Calls for diverting money from NASA's budget grew louder. The Reverend Dr. Martin Luther King Jr. addressed the issue during his testimony before the Senate subcommittee investigating the plight of inner cities on December 15, 1966. In his opening statement, King denounced the amount of money being spent on the war in Vietnam and in space:

> Beyond the advantage of diverting huge resources for constructive social goals, ending the war would give impetus to significant disarmament agreements. With the resources accruing from termination of the war, arms race, and excessive space races, the elimination of all poverty could become an immediate national reality. At present, the war on poverty is not even a battle, it is scarcely a skirmish.[20]

NASA and the civil rights movement finally met on a field just outside of the gates of Cape Kennedy. It was July 15, 1969, the day before the launch of Apollo 11 and mankind's first landing on the moon.

Led by one of the late Dr. King's closest aides, Ralph Abernathy, a group of five hundred protestors arrived at the Cape amid a light, misty rain. Abernathy had come to the forefront of the movement the prior year when he led the Poor People's March on Washington to shine a light on employment inequalities and housing issues faced by the poor across the country. On this July morning, he hoped to draw national attention once again to the plight of the 20 percent of Americans who could not afford to buy food, clothe their children, or find affordable housing.

The contrast could not be more stark. The protestors arrived with an old wooden wagon drawn by two mules symbolizing the millions of Americans in poverty. Before them stood the gleaming white Saturn V rocket that was poised to land a man on the moon. The protestors held signs that read "Moonshots Breed Malnutrition"; "Rockets or Rickets?"; and "Billions for Space, Pennies for Hunger."[21] Abernathy stood at the gates holding a sign that read, "$12 a Day to Feed an Astronaut—We Can Feed a Starving Child for $8."

Earlier that morning, NASA administrator Thomas Paine met with former president Lyndon Johnson, who questioned whether the protestors would block access to the Cape, preventing technicians from arriving for

Only days before President John F. Kennedy declared that the United States would land a man on the moon before the end of the decade, this cartoon by Thomas Stockett appeared in the newspaper the *Afro-American* on May 20, 1961, depicting the widening gap between national priorities. *Courtesy of the Afro-American Newspapers Archives*

the launch. Paine assured Johnson that he would speak with Abernathy and address the protestors' concerns. NASA attempted to keep the meeting secret, even directing Florida state troopers and NASA security to stay away to avoid drawing media attention, but news of the upcoming meeting was leaked.

At 3:00 p.m. as the protestors began singing "We Shall Overcome" and a small group of reporters gathered around, Paine made his way to meet Abernathy. Following standard government protocol, Paine wrote a Memorandum for Record to put on file describing the meeting:

The Reverend Abernathy gave a short speech in which he deplored the conditions of the nation's poor, stating that although he had no quarrel with the space program, it represented an inhuman priority at a time when so much suffering exists in the nation. One fifth of the population lacks adequate food, clothing, shelter, and medical care, he said. The money for the space program, he stated, should be spent to feed the hungry, clothe the naked, tend the sick, and house the shelterless.[22]

Paine listened intently to the reverend, soaking in every word. Finally, Abernathy made three requests of the administrator: support the movement to combat the nation's poverty, hunger, and other social problems; ask that NASA scientists and experts use their talents to tackle the problem of hunger; and allow ten families from the group inside to view the launch.

Paine granted permission for a busload of the marchers to view the launch from the VIP seating area inside the space center. When the bus arrived, Abernathy's marchers found breakfast and a candy bar awaiting each of them on the seats. In his memo, Paine wrote, "I wished it were as easy to meet his other two requests."[23]

"If we could solve the problems of poverty in the United States by not pushing the button to launch men to the moon tomorrow," he told Abernathy, "then we would not push that button. The great technological advances of NASA were child's play compared to the tremendously difficult human problems [on Earth]."[24]

Paine went on to ask Abernathy to "hitch his wagons to our rocket, using the space program as a spur to the nation to tackle problems boldly in other areas." In return, he said that he would "do everything in my personal power to help in his fight for better conditions for all Americans."

Abernathy accepted Paine's response and prayed for the three men aboard Apollo 11. "On the eve of man's noblest venture," he told reporters, "I am profoundly moved by the nation's achievements in space and the heroism of the three men embarking for the moon. What we can do for space and exploration, we demand that we do for starving people."[25]

With that, the two men shook hands and on the following day, Apollo 11 lifted off for the moon. NASA eventually modified some of its Apollo technology for use in inner cities, developing energy-efficient windows for public housing, creating new technology to measure air pollution in urban areas, and pioneering innovative water treatment technology.

The belief that the Apollo budget should be diverted to other, more pressing social issues was not felt only in America, but across the globe as well. The world watched in awe as Neil Armstrong and Buzz Aldrin prepared to take mankind's first tentative steps on another world. In London, the BBC covered the monumental event, as did most television stations across the globe. Science-fiction author Ray Bradbury sat on a panel in the BBC studios with several other guests, many of whom argued that the race to the moon was a waste of time and money. One of the panelists, Irish political activist Bernadette Devlin, finally irked Bradbury too much. The author turned to the young woman and said, "This is the result of six billion years of evolution. Tonight, we have given the lie to gravity. We have reached for the stars. . . . And you refuse [to] celebrate? To hell with you!"[26]

As the turbulent 1960s came to an end and four men had walked on the moon, the cry for diverting money from Apollo to solve the problems of the impoverished continued. Black poet and songwriter Gil Scott-Heron told his followers that the lunar missions were distractions from the country's earthly issues. "Just something to hold down the pressure and revolt in America."[27] His poem, *Whitey on the Moon*, summed up the frustrations and reality of life in urban America:

A rat done bit my sister Nell,
with Whitey on the moon.
Her face and arms began to swell,
and Whitey's on the moon.
I can't pay no doctor bill,
but Whitey's on the moon.
Ten years from now I'll be payin' still,
while Whitey's on the Moon.

But protests about the moon landings were now a moot point. Americans were completely indifferent to the program and funding for the Apollo missions had been slashed by Congress even before the first landing on the moon occurred. The budget for the moon landings dropped from $5 billion in 1965 to less than $4 billion in 1970. Follow-up programs took the brunt of the cuts. Funding for Skylab, for example, was cut from $454 million to $253 million in 1968 even before the first lunar landing.[28] As it turned out, the Apollo program was cancelled altogether in January 1970, only six months after Apollo 11. But even with the huge cuts to NASA's budget, the federal government diverted that money to other programs and issues, including the Vietnam War, but rarely to social issues.

Through those early years, quietly in the background, NASA continued making inroads desegregating its facilities and research centers across the country and fighting discrimination, sometimes at a slower pace than many would have liked due to stiff headwinds and racial attitudes, but still, progress was being made. Getting a black astronaut into space would take much longer.

6

IN MY DAY, THE BIGGEST
THING YOU COULD HAVE
DONE WAS BECOME
A SECRETARY

Jeanette and Janet Epps were born on November 3, 1970, in Syracuse, New York. The twins were the youngest of seven children born to Henry and Luberta Epps. From an early age, Jeanette was inquisitive, goal oriented, and had a dream. As she watched humans climb aboard gleaming rockets and launch into the final frontier from the sunny beaches of Florida, she knew that she wanted to be a NASA astronaut.

The young black woman was studious, earning a bachelor's degree in physics from LeMoyne College in Syracuse in 1992, then a master's in science two years later, followed by a doctorate in aerospace engineering in 2000 from the University of Maryland. Epps continually built on her education and experience, analyzing data for shape memory alloys as a NASA intern, helping research and develop automobile collision location detection and countermeasure devices for the Ford Motor Company, and even working with the CIA as a foreign weapons system analyst.

As Jeanette's career continued its journey, weaving its way around the curving bends of life, the young woman kept her eyes on the prize—to be a NASA astronaut. She eventually applied and on June 23, 2009, she received the phone call. The voice on the other end gave her the news that she had been accepted to the Astronaut Candidate Training Program. Upon her acceptance, her mother Luberta said, "In my day, the biggest thing you could have done was become a secretary."[1]

Originally scheduled to fly to the International Space Station (ISS) in 2018, Epps was reassigned and will fly on the first operational mission of the Boeing CST-100 Starliner spacecraft, which will send her to the ISS in late 2021; that would have made her the first black woman to live and work in space for an extended period of time. The groundwork that enabled this historic flight was paved by those who went before her, slowly and painstakingly over the years. As we have seen, NASA had been integrating its ground facilities and research centers as far back as 1943. That process was slow with a troubled history, but as NASA was pushing through the color barrier on the ground, sending a person of color into space proved to be much more difficult. As with many aspects of the American space program, the story of sending the first black astronaut into space began when the first human was launched into space, Russian cosmonaut Yuri Gagarin.

Until April 12, 1961, the United States was on track to launch an astronaut into space—just in its own good time. Engineers worked slowly and methodically to certify that the rockets that would hurl an American into the forbidding environment outside Earth's atmosphere were "human rated." In the wild west days of rocketry, a launch vehicle had better odds of blowing up on the launch pad than making it into space.

The same methodical approach was taken for training the first seven men who dared to risk their lives aboard what essentially was a giant stick of dynamite, with each mission leveraging the accomplishments of the previous—stepping-stones to bigger and greater space missions. That slow pace caught up with the Americans when Gagarin lifted off from the Baikonur Cosmodrome in Kazakhstan on April 12, 1961. Suddenly, the American space program was "behind," and a sense of urgency rippled through the corridors of the nation's capital.

President Kennedy was feeling pressure to jump-start the American space program, to kick it in the pants and start putting the United States ahead of the Russians and prove its technological supremacy. Less than one month later, Alan Shepard became America's first man in space, albeit a short fifteen-minute ride to the edge of space, far from Gagarin's orbital flight. In response and only three short weeks after Shepard's flight, Kennedy made the bold and audacious proclamation before a joint session of Congress that the country should land a man on the moon by the end of the decade.

The clock was ticking for NASA. Technology had to be invented. The physics of getting to the moon and the science of keeping a human alive a quarter of a million miles from Earth were only theories that now had to be put into practice. As NASA grappled with these challenges, President Kennedy grappled with other issues of national importance—basic human rights were being suffocated in much of the South. The images of the unrest and violence perpetrated against blacks were being beamed around the world, painting an indelible image that the great experiment in democracy was not living up to its promise to all Americans.

To put a shine on the nation's world image, President Kennedy invited journalist Edward R. Murrow to head the U.S. Information Agency (USIA). Murrow was the perfect choice, having established a name for himself only two decades earlier through his unvarnished reporting from the battlefields of Europe during World War II on CBS radio, making him the most trusted voice in journalism. Murrow accepted the position and immediately went to work.

Murrow recognized that these two monumental events—the battle for civil rights and the space race—were occurring concurrently but parallel to one another. Combining the two offered the country an opportunity to spotlight its technological achievements and at the same time shine a light on its progress in living up to its promise that all men and women are equal. He quickly typed out a note and sent it to NASA Administrator James Webb.

> Dear Jim,
>   Why don't we put the first non-white man in space? If your boys were to enroll and train a qualified Negro and then fly him in whatever vehicle is available, we could retell our whole space effort to the whole non-white world, which is most of it.[2]

Webb's response was predictable and avoided answering the request. It was the standard reply: "our selection process is based on technical qualifications and requirements and only qualified applicants will be considered without regard to race, color, or creed."[3]

By now, Kennedy was feeling the pressure to not only successfully land on the moon, but also to integrate the astronaut corps. The White House was flooded with letters and telegrams asking the president to consider the idea, even one from a "proud Southerner," marketing specialist Robertson McDonald:

I have always been a proud American and a proud Southerner. The successful orbiting of Col. John Glenn causes this great pride to be even greater.
However, I am somewhat disturbed at the obvious discrimination as Commander in Chief of the Armed Forces that you have shown towards the
American Negro. Why isn't there a Negro astronaut?[4]

Kennedy decided to find out whether any black pilots were being considered for spaceflight and tasked his close personal aide, Fred Dutton, to get
the ball rolling. Dutton fired off a memo to Adam Yarmolinsky at the Pentagon. During his career, Yarmolinsky became known as Secretary of Defense
Robert McNamara's "whiz kid," modernizing the Pentagon's complex man-
agement systems and later helping to draft the Gesell Report, which would
end discrimination in the military. He went on to assist Lyndon Johnson in
drafting the president's plans for the War on Poverty, for which columnist
Robert Novak called Yarmolinsky the "chief midwife in the hurried birth of
the poverty program."[5]

In his memo, Dutton wrote, "I would appreciate knowing whether any
of the fifty Air Force test pilots who have been selected for future military
men-in-space missions, as announced in the *New York Times*, Sunday, July
23, are from minority groups."[6]

Yarmolinsky deftly skirted the question and gave the same form-letter
response as Webb: "Specific criteria for the selection of personnel for future
space endeavors has not yet been resolved. The criteria . . . will not exclude
from participation any individual of any minority group."[7]

Dutton fired back a reply, emphasizing that it was desirable to "include
representatives of minority groups in significant undertakings." He went on
to demand a report detailing the recruitment and selection process of minority astronauts and when the administration could expect a minority to be
selected. Dutton set a firm date for a reply—November 1, 1961.[8]

There was indeed a black test pilot in the Air Force who was making his
way up the ranks and who would be a likely candidate to be the first black
astronaut. His name was Edward Joseph Dwight Jr.

Born in Kansas City, Kansas, in 1933, Ed Dwight had natural artistic
ability and took to drawing at an early age, completing his first oil painting
at the age of eight. Besides art, something else piqued his interest—flying. At
the age of six, Dwight could be found at the local airport cleaning airplanes,
and by the age of ten, he wanted to fly but never dreamed it actually would

happen. He had seen racism and segregation firsthand when his mother enrolled him in a private Catholic school. According to Dwight, of the 850 students enrolled in the school, 300 dropped out when he arrived.[9] His feeling about being allowed to fly changed, however, when he saw a photo of a black fighter pilot who had flown missions in Korea. Dwight thought to himself, "Oh my God! They're letting Black folks fly jets!"[10]

Dwight went on to college and earned an associate degree in engineering before joining the Air Force, where he became an ace fighter pilot, racking up 2,200 hours of flight time, 1,700 of those in high-performance jets. He went on to train other pilots in instrumentation but was told that he was not eligible for squadron leader because "country boys wouldn't want to follow [him]."[11] But that didn't deter the young man whose determination, experience, and education landed him at the elite Air Force test pilot training school at Edwards Air Force Base in California to begin phase-one pilot training, where he possibly could progress through the ranks and apply to NASA to become an astronaut. His eight-month-long training would be completed under the guidance of the man who first broke the sound barrier, Chuck Yeager.

Dwight recalls that upon his arrival at Edwards, a senior officer questioned his qualifications. "Why in the hell would a colored guy want to go into space, anyway?" Dwight recalls him saying. "As far as I'm concerned, there'll never be one to do that. And if it was up to me, you guys wouldn't even get a chance to wear an Air Force uniform."[12]

Even Dwight's hero, Chuck Yeager, was not keen on the pilot joining the program. In his memoir, *Yeager: An Autobiography*, the pilot wrote that Dwight was an average pilot with an average background and that Dwight required extra tutoring to make it through training,[13] to which Dwight emphatically replies to this day, "That is absolutely absurd!"[14]

Despite the strong headwinds of racism Captain Dwight was fighting, he proved the skeptics wrong and completed the first phase of training in April 1963 along with fourteen other candidates, all of whom were white. It was on to phase two, aerospace research pilot training. When word that Dwight had completed phase-one training and was one step closer to becoming an astronaut was announced, the media started to take notice. Headlines in newspapers around the country shouted the exciting news: "Astronaut's Parents So Excited" and "Negro Astronaut Aiming for the Moon." They speculated that the pilot could become the first black astronaut, maybe even be the first man to walk on the moon. Even the *New York Times* took notice with a headline reading, "Negro One of 15 in Space Course."[15] The captain became something of a celebrity, receiving more than fifteen hundred pieces of mail each week from fans and well-wishers.

Dwight went on to graduate the phase-two training and was recommended to NASA for a shot at becoming the first black astronaut.

Out in the deserts of Nevada, the chief of the Astronaut Office and one of the original seven astronauts selected for Project Mercury, Deke Slayton, was training a group of potential candidates to fly in the follow-up to NASA's Mercury missions, Project Gemini, when he was interrupted by a phone call. When he returned to the trainees, he made an announcement. He told them that he had just finished a telephone conversation with Attorney General Robert Kennedy, who strongly suggested Slayton accept Dwight into the new group of astronauts. Slayton told the men, "I just spoke for all of you guys. I said if we had to take him and he wasn't qualified, then they'd have to find sixteen other people, because all of us would leave."[16]

In October 1963, the media gathered inside a conference hall and was introduced to the new Gemini astronauts. Ed Dwight was not one of them. A journalist directed a straightforward question at Slayton: "Was there a Negro boy in the last 30 or so that you brought here considered?" The chief of the Astronaut Office replied with a short, stony, and firm, "No. There was not."[17]

Outrage ensued throughout black media. *Ebony* and *Jet* magazines and the *Chicago Tribune* ran articles about NASA dropping the first black astronaut candidate. The National Association for the Advancement of Colored People (NAACP) demanded that the Department of Defense investigate his dismissal. On the other side of the world, the Russians seized upon NASA's decision to exclude Dwight from the astronaut corps and used it for propaganda purposes. Once again, Soviet headlines blasted the Americans for their hypocrisy on civil rights, the pages brandishing images of the violence that blacks in America were suffering in the South and linking that turmoil with this space age snub.[18]

Following NASA's announcement of the new astronauts, Dwight was met by a group of reporters as he stepped off a plane. In newsreel footage,

Dwight is visibly shaken but he handled the moment with dignity, refusing to comment or lay blame as to why he was not selected. He quietly retired in 1966, returning to pursue his first love, art. Today, Dwight is a sculptor who has created thousands of gallery pieces and hundreds of memorials to notable black figures and events in history including the Martin Luther King Jr. Memorial in Denver and the Underground Railroad monument in Detroit. Today, whenever asked about almost becoming the first black astronaut and the fame that brought, he replies, "How the hell do you get famous for something you've never done?"[19]

Outside of NASA, the Air Force and the National Reconnaissance Office began work on a secret space project—the Manned Orbiting Laboratory, or MOL. Work on MOL was authorized by President Johnson in 1965 and would consist of a series of small space stations or laboratories, which were basically empty, pressurized Titan rockets (the same rockets that boosted Gemini spacecraft into orbit) with an attached capsule that looked remarkably like the Gemini. These stations would orbit Earth and be manned by a two-man crew that would fly thirty-day missions high above America's Cold War enemies, taking high-resolution images of their missile launch sites as well as their nuclear and military facilities.[20]

The men selected to man these space stations included future shuttle pilots Robert Crippen, Gordon Fullerton, and Richard Truly (who would later become NASA's first astronaut administrator). The team also included a young black Air Force pilot, Major Robert H. Lawrence.

The Chicago native joined the Air Force at the age of twenty immediately after earning a bachelor's degree in chemistry from Bradley University. The pilot had a stellar flying career, logging more than 2,500 hours in 2,000 jets. While in the Air Force, he earned a PhD in physical chemistry from Ohio State University.[21] After completing test pilot training, his commanding officers recommended Lawrence to the MOL program, and training for this unique mission began.

On December 8, 1967, Lawrence sat in the rear seat of an F-104 Starfighter supersonic jet with Major Harvey Royer piloting in the front seat. The goal of this training session was to practice landing techniques that would eventually be used to land the space shuttle—climbing to 25,000 feet and performing a 25-degree dive at 330 miles per hour. On

Third group of Manned Orbiting Laboratory astronauts: Robert T. Herres, Robert H. Lawrence, Donald H. Peterson, and James A. Abrahamson. *NASA*

the jet's final approach, the aircraft's landing gear hit the runway and collapsed, causing the plane to go airborne briefly before crashing down on the tarmac. With its fuselage engulfed in flames, the aircraft veered off the runway and began to disintegrate. Both pilots ejected. Royer was injured but survived. Lawrence's chute, however, did not deploy. He was killed in the accident.[22]

Lawrence left behind his wife Barbara and eight-year-old son Tracey. As if losing a devoted husband and loving father wasn't enough, his wife suffered through the pain of racism. Although she received many letters and telegrams expressing grief and support during her time of need, she also faced vile racism. One letter she received is etched vividly in her memory. It read, "Glad he was dead because there would be no coons on the moon."[23]

As the 1960s faded into the 1970s, the Nixon administration decided that drastic cuts to NASA's budget were needed. Piloted spaceflights would be curtailed in favor of robotic missions to the planets. NASA's goal of landing humans on Mars by 1981 was scrapped and the final two

Apollo moon landings—Apollo 18 and 19—were cancelled. The remaining Apollo hardware would be used for the Apollo-Soyuz Test Project (ASTP), in which American and Russian astronauts would demonstrate that the two superpowers could cooperate in space and make three flights to a rudimentary space station known as Skylab.[24]

The budget did, however, retain one piloted component: the development of a reusable spacecraft to be known as the space shuttle, a workhorse of a vehicle that would launch and repair satellites from Earth's orbit, from which an orbiting space station would be built, and that would ferry astronauts to the station to live and work. And, of course, this new project required an entirely new crop of astronauts.

In January 1978, the media once again converged upon a conference room in Houston, Texas, where they were introduced to a new group of astronauts who would fly the shuttle and build the space station. It was the first such introduction in nearly ten years. Known as the "thirty-five new guys," the difference between this group and previous ones could not have been more striking. Of the thirty-five, the class of 1978 included the first Asian American, six women, and three black recruits—Guion "Guy" Bluford, Frederick Gregory, and Ron McNair.

Guy Bluford knew from an early age that soaring high in the sky was in his future. He loved to build model airplanes as a child and attended college at Penn State, where he graduated as a distinguished Air Force ROTC grad with a bachelor's degree in aerospace engineering. Upon graduation, he joined the Air Force, where he served as a pilot during the Vietnam War, flying 144 missions, 65 of those over North Vietnam. He eventually racked up more than 5,200 hours of flight time during his military career. After earning his master's degree in aerospace engineering in 1974, he went on to earn a doctor of philosophy degree in aerospace engineering from the Air Force Institute of Technology.

All of this education and experience led the young Air Force colonel to NASA's new space shuttle astronaut corps. But simply being selected to join the team didn't mean you had a ticket to ride. The road to being selected for a mission was a long and arduous one.

To hitch a ride on the space shuttle required candidates to go through one of the most rigorous training processes in the world. Even before candidates were selected to fly specific missions, they first had to undergo two years of intensive, basic training, "boot camp," if you will, on shuttle systems, space science, engineering, earth science, meteorology, land and sea survival, scuba diving, and aircraft operations. Once assigned to a mission, their specialized training would begin.

While Bluford, McNair, and Gregory began their journey to become shuttle astronauts, on the other side of the world, the Russians had one more space first up their sleeve. It would be the culmination of their propaganda efforts lambasting America for its civil rights failures. They would be the first to send a black astronaut into space.

It began in 1976 with a joint agreement among the Soviet Union and its allied nations of Cuba, Czechoslovakia, Poland, East Germany, and Romania that would allow citizens of those countries to fly manned missions into space on Soviet rockets. Rather than going through the rigors of flight engineer training, this new breed of cosmonaut instead would train in a slimmed-down version of the program and be known as "cosmonaut-researchers," the Russian version of the space shuttle's "mission specialist."[25]

The first three flights of this new program called Intercosmos would occur in 1978 and were reserved for candidates from Poland, Czechoslovakia, and East Germany, the three nations that had helped with the development of the equipment that was to be used in the program. The following year, the program included trainees from other nations, including Cuba. One of those candidates was Brigadier General Arnaldo Mendez.

A Cuban of African descent, Mendez was orphaned at an early age but eventually adopted by his foster parents, Rafael and Esperanza Mendez. He began work at the age of thirteen shining shoes in his hometown of Guantanamo. Mendez first took to the air after completing aviation courses in Cuba in 1960 and then studied at the Yeisk Higher Air Force School in the Soviet Union where he learned to fly MIG fighter jets. This training led to the young pilot flying twenty reconnaissance missions over his home island and the southern United States during the Cuban Missile Crisis of 1962.

In 1978, Mendez applied to train for the Intercosmos program. After a rigorous process, he was selected and began two-and-a-half years of training at Star City. On September 18, 1980, Mendez and cosmonaut Yuri Romanenko boarded the Soyuz 38 spacecraft and launched into the blue Kazakhstan sky, making Mendez the first black man in space. The mission lasted twenty hours and forty-three minutes, during which time the space-craft docked with the Salyut 6 space station and deployed and carried out thirty-seven experiments before returning safely to Earth.

Back in the United States, a crew was being selected for the eighth flight of the space shuttle program. One of NASA's three black trainees would be aboard *Challenger* for the mission. Guy Bluford never intended to be America's first black astronaut. In his words, "My desire was to make a contribution to the program."[26]

"All of us knew that one of us would step into that role," he said. "I probably told people that I would probably prefer not being in that role . . . because I figured being the number-two guy would probably be more fun."[27]

With a long list of accomplishments and experience under his belt, Bluford was assigned to fly aboard STS-8. On August 30, 1980, with a light misty rain falling, the shuttle roared to life in the early morning hours and rose from its launch pad on the Florida coast with thousands of onlookers and VIPs cheering Bluford and the crew on. Those in attendance were not the only ones who experienced the sheer excitement of the launch. All communications between the spacecraft and ground controllers are recorded, and during playback, engineers were surprised by what they heard. "I laughed and giggled all the way up," Bluford recalls. "It was such a fun ride."[28]

Bluford went on to fly three additional shuttle missions, logging 688 hours in space. Frederick Gregory flew on three shuttle missions, logging 455 hours in space and eventually was confirmed as NASA's deputy administrator in 2005.

Ron McNair saw his dream of becoming an astronaut come true when he flew aboard the shuttle *Challenger* on STS-41B, which launched on February 3, 1984. Tragically, McNair's ambitions were cut short on January 28, 1986, when that same shuttle was consumed by the explosion of the rocket's fuel tank, killing McNair and his fellow astronauts Francis "Dick" Scobee, Ellison Onizuka, Christa McAuliffe, Gregory Jarvis, Michael Smith, and Judith Resnik.

Through 2021, fourteen black American astronauts have flown into space, which leads us back to Jeannette Epps and her historic flight to the ISS aboard the Boeing Starliner. However, the story of black astronauts is far from over. On December 9, 2020, NASA introduced the eighteen

Stephanie Wilson (bottom left) is pictured with (counterclockwise from bottom right) crewmates Dorothy Metcalf-Lindenburger, Naoko Yamazaki, and Tracy Caldwell Dyson in the cupola of the International Space Station. *NASA*

astronauts of the Artemis project that will return humans to the moon. Of the eighteen, three are black—space station veterans Stephanie Wilson and Victor Glover as well as geologist Jessica Watkins. All three are in line for another possible space first—the first black astronaut on the moon.

7

THE PROBABILITY OF
SUCCESS IS DIFFICULT
TO ESTIMATE

*Fate has ordained that the men who went to the moon to explore
in peace will stay on the moon to rest in peace. These brave men,
Neil Armstrong and Edwin Aldrin, know that there is no hope for
their recovery. But they also know that there is hope for mankind
in their sacrifice. For every human being who looks up at the moon
in the nights to come will know that there is some corner of another
world that is forever mankind.*[1]

In 1999, while researching the origins of America's relationship with
China during the Nixon administration, historian James Mann found
himself rummaging through the darker recesses of the Nixon archives
when he made an astonishing discovery. Buried deep within the otherwise
meticulously organized records was a two-page memo from one of Nixon's
speechwriters, future Pulitzer Prize–winning journalist and Presidential
Medal of Freedom recipient William Safire. The yellowed document was
addressed to the former president's chief of staff, H. R. Haldeman. Titled
"In the Event of Moon Disaster" and dated July 18, 1969 (only two days
before the historic first landing on the moon by the crew of Apollo 11), the
memo was a what-if speech that the president would read in the event that
Neil Armstrong and Buzz Aldrin landed on the moon but were unable to
return. Stranded on the lunar surface, the astronauts would either die of
asphyxiation or commit suicide via cyanide.

Page two of the memo held specific instructions: Prior to the president's statement, he should telephone each "widow to be." After the president's statement, NASA would end communications with the men (what Safire called a euphemism for "committing suicide")[2] and a clergyman would perform the ceremony used for burials at sea, commending the astronauts' souls to "the deepest of the deep" then concluding with the Lord's Prayer.

For the thirtieth anniversary of the Apollo 11 moon landing, Mann penned an article for the *Los Angeles Times* announcing his discovery. One of those who read the article was William Safire himself, who was immediately drawn back to memories about how the document came about in 1969. In an interview with the *New York Times*, Safire recalled that he was approached by Gemini astronaut and Apollo 8 alumni Frank Borman to write the contingency speech. At first, Safire didn't quite understand what the pilot was asking him to do.

"You want to be thinking of some alternative posture for the president in the event of mishaps," the astronaut said. Safire admits he was unsure of what Borman was asking for until it was made perfectly clear: "Like, what to do for the widows."[3]

The discovery of the memo laid bare the undeniable truth that manned spaceflight was a risky adventure with unpredictable consequences. As MIT professor Phillip Morrison once said about the advancement of science and knowledge in general, "the probability of success is difficult to estimate."[4]

Thankfully, Nixon never had to make that speech, and it was relegated to the dusty back shelves of the archive. However, the potential for disaster in spaceflight is ever present although most people rarely give it a second thought. Aside from incredible space firsts like the first manned flights by Yuri Gagarin and Alan Shepard or the first lunar landing, spaceflights have become dull and routine to the public. Launches to the International Space Station are viewed with a ho-hum attitude, almost like jumping into a car and heading off to work. To them, flying into space is just another commute to work. That perception is not the public's fault. NASA and the many space agencies around the world make spaceflight look easy, so naturally the public becomes blissfully ignorant of the inherent dangers involved when an astronaut is strapped into a tiny capsule on top of what effectively can be described as giant stick of dynamite. Every now and then, the world is tragically snapped back to reality—the Apollo 1 flash fire, the *Challenger* explosion, the disintegration of *Columbia*. But as soon as the next flight lifts off from the sunny Florida coast, like the Phoenix rising from the ashes, the journey once again becomes routine.

There are many tales in the annals of mankind's exploration of space that show just how dangerous this endeavor is, events that range from mere nuisances to near disasters that never made the headlines but nevertheless occurred.

One of the first manned spaceflights to encounter mechanical issues that could have led to a disastrous ending was the flight of MA-6 (Mercury-Atlas 6) and *Friendship 7* with Colonel John Glenn onboard. The flight was to be a test of how well a human could perform in outer space but what it is most remembered for is that it was the first time an American astronaut orbited Earth.

Settling into orbit after a picture-perfect launch, Glenn relaxed to enjoy the ride while performing a list of tasks and experiments. As the spacecraft floated effortlessly through space, Glenn began to notice that it was drifting off course a degree or two to the right, much like a car out of alignment. At about the same time, the tracking station in Guaymas, Mexico, relayed a message to mission control in Florida that *Friendship 7*'s automatic stabilization and control system was causing the issue. Basically, the spaceship's thrusters were firing, causing the capsule to drift.

After switching back and forth several times from the automatic "fly-by-wire" system to manual piloting, Glenn realized that he would have to pilot the craft manually and try to conserve fuel so that he would have enough for the maneuvers required to line him up for reentry and the return to Earth.

Glenn acclimated to flying in space quite well and the mission continued without incident—that is, until William Saunders, who was manning a telemetry monitoring console on the ground, noted strange readings coming from the spacecraft's landing system. The Mercury capsule was fitted with a set of retrorockets that would slow the capsule to begin its return to Earth. These rockets were connected to the underside of the capsule by straps that would be jettisoned after the rockets had done their job. Beneath the straps lay the all-important heat shield that would protect the capsule from the searing 3,000-plus degree heat of reentry and underneath that, a landing bag that would deploy and cushion the capsule's plunge into the ocean. What the data told the engineer was that the heat shield and landing bag was not locked in position and was held in place only by the straps of the retro-rockets.[5]

A decision had to be made: should they keep the retro-rockets attached to the spacecraft throughout the reentry process to ensure the heat shield stayed in place or continue with reentry as planned by jettisoning the retros and possibly jeopardizing Glenn's life? Could they rely on the data that was coming in? After consulting with the capsule's designer, Maxime A. Faget, it was decided to leave the rockets strapped to the capsule just to be safe. It was a decision that would leave the flight controllers with a palpable uneasy feeling for the remainder of the flight.

Until his third and final orbit, Glenn was blissfully unaware of the potential danger he faced. He knew something was afoot after being questioned multiple times about the status of the landing bag deploy switch. It wasn't until flight director Chris Kraft sent the order for Glenn to place the switch in the automatic position that he fully understood what was going on. His orders were to leave the retro pack in place if a light came on.

Glenn radioed mission control that he would be manually flying the spacecraft during reentry. The retro-rockets did their job, slowing the spacecraft and allowing it to begin its descent. Controllers told Glenn to keep the retros attached until the spacecraft entered the atmosphere and experienced 1.5Gs. When indicators showed that the capsule was experiencing 0.5G, Glenn pushed the override button. As soon as he did, he began to hear noises that sounded like something brushing against the capsule.

"That's a real fireball outside," he radioed with a bit of anxiety in his voice.[6] He had good reason to feel that way. At that moment, one of the straps had broken loose and the smoking strip of metal sailed past his window before being consumed by fire. Burning chunks of metal flew past the window, and he feared that the heat shield was gone.

As Glenn reached maximum g-forces, the capsule began gyrating to the point that the astronaut could no longer control it. He activated auxiliary systems to smooth out the ride, but soon after, the control thruster's fuel supply was exhausted. "I felt like a falling leaf," he later commented.[7]

Soon, the spacecraft was in the proper position for parachute deployment. The chutes unfurled, the landing bag deployed, and with a reassuring jolt, *Friendship 7* landed safely in the Atlantic Ocean slightly more than four hours after launch and three orbits around Earth.

No matter how you look at it, being an astronaut is a dangerous profession. From liftoff to splashdown and everything in between, peril lies around every corner. Astronauts enter their spacecraft with the knowledge that engineers have thought through as many scenarios as possible and have devised multiple contingency plans to deal with whatever crops up. Even with that comforting knowledge, it still takes a special kind of person to hop aboard a rocket knowing the countless dangers, caveats, and what-ifs. To have the wherewithal to stay calm and cool under immense pressure should something go wrong and to save the mission—and their lives—is a testament to the courage of these men and women. Astronaut Wally Schirra demonstrated such qualities during the launch of Gemini 6A in 1965.

In October of that year, NASA was preparing for back-to-back space firsts. Gemini 6 with veteran Mercury astronaut Wally Schirra and rookie Tom Stafford aboard would perform the first docking in space with another vehicle. Shortly after that mission had landed safely, Gemini 7 would be launched with Frank Borman and Jim Lovell aboard who would break the spaceflight duration record.

The vehicle Schirra and Stafford were to dock with was called an Agena. The long, cylindrical spacecraft was an upper stage booster rocket developed by the Lockheed Corporation, which would prove to be a workhorse for NASA and the Air Force. Since 1959, 150 Agenas had been launched, lofting communications and reconnaissance satellites into orbit. This would be the first time that the Agena would be modified so that the astronauts in the Gemini capsule could physically dock with it, a key step if the Americans wanted to eventually land a man on the moon.

On October 25, the countdown began for the 10:00 a.m. launch of the Agena. The plan was to launch the Agena first, followed by Gemini 6 a few minutes later. At T-minus fifteen minutes, Schirra and Stafford made their way to the Gemini launch pad and were strapped into the capsule. Right on time, the Agena rose from Launch Pad 14 into the blue Florida sky. Public affairs officer Paul Haney relayed to the media and to the public watching on television and at Cape Kennedy that the launch was flawless and the vehicle was working nominally. Minutes later, that all changed. Tracking stations reported that they had lost all telemetry from the Agena. Air Force radar stations reported that the vehicle appeared to have exploded and it was tracking pieces of it plummeting to Earth. Not long after, Haney reported that the

Gemini VI commander Walter Schirra (seated) and pilot Thomas Stafford perform suiting-up exercises in preparation for their forthcoming flight in this image from October 1965. The suit technicians are James Garrepy (left) and Joe Schmitt. On the third attempt, Gemini VIA successfully launched on December 15, 1965. *NASA*

vehicle had been lost. "There was no joy. No joy," he said.[8] With the loss of the Agena target vehicle, the flight of Gemini 6 was scrubbed.

NASA was in a quandary regarding what to do with the Gemini 6 mission since there was no target vehicle to dock with. Engineers John Yardley and Walter Burke had the answer. Gemini 7 would proceed with its launch in December. Gemini 6 (now renamed Gemini 6A) would launch soon after and chase Gemini 7 to perform a close-up rendezvous with another vehicle in space. The Soviet Union had successfully performed such a rendezvous two years earlier with the pioneering flight of Valentina Tereshkova during the Vostok 5 and Vostok 6 missions, but the two spacecraft only came within three miles of one another. If Gemini 6 and 7 were successful, they would meet nose-to-nose only a few feet apart.

At 2:30 p.m. on December 4, Gemini 7 rose from its launch pad and entered Earth orbit. Launch pad crews wasted no time and began preparing the pad for the launch of Gemini 6A. Seven days later, the Titan rocket that would propel Schirra and Stafford to their rendezvous with Gemini 7 was ready. The countdown was flawless and at 9:54 a.m., the Titan's turbo pumps screamed their telltale high-pitched whine as they revved up and the rocket's dual engines roared to life. A thick, billowing orange cloud engulfed the base of the rocket. Launch controllers announced, "lift off!" Onboard the capsule and at mission control in Houston, the mission clocks began ticking away, indicating that the flight had officially started. Mission control bellowed over the communications circuit, "the clock has started!"

Normally, the commander of a flight acknowledged the call that the clock has started by repeating it. This time, however, Schirra didn't answer.

"I knew we hadn't lifted off," Schirra later wrote. "It was a gut feeling. Stafford didn't know what was going on, but I had the experience of a Mercury flight and my butt told me we hadn't left the pad."[9]

The clock had indeed started, but just as quickly, the rocket's engines had shut down without moving an inch. That orange cloud of rocket exhaust now took on a menacing appearance. Onboard the capsule, the two astronauts had to make a split-second decision.

"Our choices were very limited," Schirra said. "We could stay there and blow up, we could stay there and be safe, or we could eject and that would be the end of the mission."[10]

Unlike the previous Mercury missions and later Apollo flights that had escape towers attached to the top of the capsules that propelled the astronauts to safety in the event of a launchpad emergency, Gemini astronauts had to eject from the capsule, a dangerous and unproven event. The pair would rocket out of the capsule almost horizontally at an incredible velocity in the hope that parachutes would land them safely on the ground away from the explosion.

"All the electronic circuits said we lifted off," Schirra continued. "If that had in fact happened and we settled back [to the ground], we would have blown up. The thing to do by mission rules was to eject. But I knew in my soul, in my body, that we had not lifted off and Tom Stafford said, 'OK, Wally, I buy it.' We didn't eject. [Stafford] sat there breathing rather heavily."[11]

Reviewing the incident, engineers discovered that a single plug had fallen out of the engine causing the abrupt shutdown. The rocket's systems performed as they were designed and prevented liftoff and an explosion. Schirra has been credited with having ice water in his veins during what NASA described as one of the most suspense-filled moments of the Gemini project. His actions saved the mission, which was readied for a second launch attempt with incredible speed and successfully launched a few days later. A few hours after launch, the two spacecraft rendezvoused and floated together only one foot apart. In response to the adulation, Schirra said that if the same launch failure had occurred on the second launch attempt, there was no doubt he would have pulled the ejection ring.

The launch of Gemini 6A was the first such engine shutdown for the American space program with astronauts onboard, but it wouldn't be the last. During a few of those launches, the spacecraft actually lifted off before problems occurred. The space shuttle fleet experienced engine shutdowns a few times during its run from 1981 to 2012, the most tragic being the explosion of the shuttle *Challenger*'s fuel tank only seconds after launch, killing all seven astronauts. But that wasn't the first time that *Challenger* had engine issues. Only six months prior to the tragedy, the same shuttle came close to plummeting back to Earth after an engine failure, but quick thinking by mission control engineers averted disaster.

After a ninety-seven-minute delay, STS-51F launched from Cape Kennedy at 5:00 p.m. on July 29, 1985. Two minutes into the flight, the solid rocket boosters dropped away, and the shuttle's three main engines continued powering the craft skyward. Three minutes and forty seconds into the flight, a temperature sensor on a fuel pump turbine detected that it had reached a temperature of 1,960 degrees, causing the sensor to shut down. A backup sensor came online and quickly rose to 1,850 degrees, which caused *Challenger*'s computers to shut down its number-one main engine.[12]

At this point, the shuttle was past the point of aborting the mission and making a safe landing in Spain, but it was not traveling fast enough to reach orbit. Mission control radioed to the crew to "ATO" (abort to orbit). The call caused the crowd watching the launch from Cape Kennedy to utter an audible gasp and created a panic in the press pool.[13] To help boost the shuttle's speed, quick-thinking engineers had the astronauts fire the shuttle's orbital maneuvering system, or OMS rockets. These are small rockets located on the back of the shuttle just above and to the sides of the shuttle's three main engines. They provide mobility for the spacecraft while in orbit. Engineers told the crew to override the sensor that told the shuttle's computer to shut down one main engine and ignite the OMS. This sequence burned off fuel and lightened the shuttle by 4,400 pounds. This quick thinking provided enough velocity to put the crew into orbit.

NASA engineer Cleon Lacefield addressed reporters after the shuttle was safely in orbit. "If this had not been done," he said, "and if the single remaining sensor on the right engine had reached its limits, the computers would have automatically shut down the second engine, causing *Challenger* to fall to Earth. If the right engine failed . . . we would have been in the water."[14]

The Russian space program was not free from similar launch malfunctions. One hair-raising moment occurred on April 5, 1975, when cosmonauts Vasili Lazarev and Oleg Makarov boarded the Soyuz 18 spacecraft for a scheduled two-month stay aboard the Salyut 4 space station. It was a flawless launch until almost five minutes into the flight. At an altitude of ninety miles, the first stage booster was supposed to separate, and the second stage engine would ignite, hurling the cosmonauts into orbit. On this flight, however, only half of the pyrotechnic charges ignited, leaving the first stage partially attached to the second. The second stage ignited, dragging the spent booster behind it.

The rocket began what the Soviets reported as "unplanned motions"[15] and the flight had to be aborted. At this altitude, the launch escape system already had been jettisoned, which forced mission controllers to manually initiate the separation of the third stage and, ultimately, the capsule from the ground. Normally, astronauts and cosmonauts experience only three to four times the force of gravity (or Gs) on reentry, but this high-altitude maneuver sent the cosmonauts into a ballistic dive that, for a time, resulted in a plummet to Earth of 3.5 miles per second, reaching maximum g-forces of more than 20Gs.

The capsule's parachute successfully deployed, and the crew landed safely on a snow-covered slope 978 miles from its launch pad in western Siberia. Although the capsule had made a soft landing, the wild ride wasn't over. The spherical capsule began rolling downhill, heading ever closer to a sheer rock cliff.[16] The only thing that prevented the cosmonauts from meeting a horrible end at the bottom of that precipice was the capsule's parachutes, which snagged a tree and prevented disaster. Both cosmonauts survived the ordeal. Lazarev suffered internal injuries and never flew again.[17]

With the exception of the Gemini and space shuttle programs in the United States and Vostok in Russia, all manned rockets have been equipped with launch escape systems (LES). An LES is a tower attached to the top of a capsule equipped with a series of high-powered rockets. These rockets could be fired either automatically or manually in the event of a booster disaster, blasting the capsule and its occupants safely away from the launch pad or booster. As a note, the new SpaceX Crew Dragon capsule uses a similar system, except the rockets are built into the capsule and not on a tower.

Until 1983, neither country had to use the LES—that is, until the launch of Soyuz T-10A, which proved that the science behind LES systems was sound. However, the unintentional test nearly exacted a high price.

By September 26, 1983, the Russian R-7 booster had successfully launched cosmonauts into space ninety-four times, so it is understandable that the crew and engineers prepping Soyuz T-10A for launch were quite relaxed as they performed their final preflight checklists. Vladimir Titov and Praskovya Mikhailovna Strekalova had been seated inside the capsule for three and a half hours, listening to music as they checked switches and reviewed flight plans before their scheduled liftoff just after midnight local time. Their goal was to dock with the Salyut 7 space station and begin a long-duration mission. Earlier that morning, Titov had spoken with his mother, who begged him not to fly the mission.

The countdown continued flawlessly until ninety seconds before liftoff, when a valve that supplied fuel to the booster failed to close, and rocket fuel poured onto the launch pad. At T-minus thirty seconds, the fuel ignited and flames began consuming the booster. In a safety bunker far from the pad, launch director Aleksei Shumilin spotted the flames and anxiously waited to see the launch escape system shoot the cosmonauts to safety. Nothing happened. The fire was so intense and spread so rapidly that it had burned through the wiring that would initiate the automatic escape.

Luckily, there was a backup system: two engineers in separate buildings near the pad would have to push the abort button at precisely the same moment. Shumilin sent the code word to the engineers, "Dnestr!"[18] (the name of a river that runs through Ukraine). With only seconds to spare, the explosives fired, and the capsule shot like a bullet into the dark Baikonur sky just as the booster was swallowed by a hellish orange-and-red fireball.

The cosmonauts suffered only minor bruises when the spacecraft landed two and a half miles away from the explosion. Upon reaching the pair, the medical team gave each of them a glass of vodka to calm their nerves.[19]

As always, the Russians didn't disclose how serious the accident was. Reports over Russia's international shortwave radio station, Radio Moscow, announced to the world that its space program suffered an accident but quickly shifted gears to boast about the success of its launch escape system. It wasn't until three years after the *Challenger* disaster in 1986 that the Russians acknowledged the explosion, saying that they were "six seconds from a Soviet *Challenger*."[20]

As dangerous as launches are, landings are just as perilous. A multitude of systems must work precisely and in the correct order for a spacecraft to make it through Earth's atmosphere safely. First, the retro-rockets have to fire at precisely the right moment and burn for the correct amount of time to begin the descent. If the spacecraft reenters at an angle too steep, it will burn up; too shallow, and it skips off into space never to be heard from again. The parachutes must deploy at the correct altitude and fully unfurl. Not to mention the million other components that have to work exactly as designed. The American astronauts selected to take part in the first joint US–Soviet Union spaceflight found out the hard way what happens when one of those elements fails.

The mission began in 1972 when, at the height of the Cold War, President Richard Nixon and Soviet Premier Aleksey Kosygin signed a joint-cooperation agreement that would see the two space superpowers join forces for the peaceful exploration of space. The first mission would demonstrate the viability of working together, prove the technology that would be required for docking together two spacecraft with completely different systems, and, more importantly for the leaders of the two nations, provide even greater prestige in the world. The flight would be called the Apollo-Soyuz Test Project, or ASTP. Naturally in Russia, the Soviets had to be first and called it the Soyuz-Apollo Test Project.

The Russian capsule was labeled Soyuz 19, but the American Apollo capsule was not assigned a name or number as with previous Apollo vehicles. It was a leftover from the cancelled moon landing program and would be the last Apollo capsule to fly. The Russian crew would include the first man to walk in space, Alexey Leonov, and veteran cosmonaut Valery Kubasov. For the Americans, the obvious choice for commander was Tom Stafford. Stafford had more experience rendezvousing with spacecraft than any other astronaut or cosmonaut, having piloted Gemini 6A to within a foot of Gemini 7 in 1965, docked with an Agena target vehicle with Gemini 9, and docked with the lunar module while orbiting the moon aboard Apollo 10 in a test run before the first lunar landing. Also on board were rookies Vance Brand and Donald "Deke" Slayton. Slayton was one of the original Mercury 7 astronauts who had been grounded due to a heart condition and never flew a mission. He had been relegated to heading the astronaut office from Mercury through the Skylab missions of the early 1970s. When the ASTP flight was developed and assignments were announced, Slayton's health was finally given the green light and he was assigned to the mission.

The flight went off without a hitch and on July 17, 1975, the two vehicles docked together using a special adapter to accommodate both craft. At 3:17 p.m., the world witnessed something extraordinary—the hatch in the docking module swung open and astronauts and cosmonauts shook hands hundreds of miles above Earth, officially ending the Cold War in space.

After exchanging gifts, performing rudimentary experiments, enjoying dinner together, and signing the official mission certificate, the hatch was closed and the Apollo spacecraft pulled away from the Soyuz, ending the joint portion of the mission. Both spacecraft remained in orbit to conduct further experiments on their own with the Soyuz returning to Earth on July

21. Apollo remained in orbit an additional three days before beginning the reentry process.

When a capsule returns to Earth, it becomes a blazing fireball resembling a meteor plunging through the atmosphere. The temperature outside the craft heats up to more than 2,600 degrees.[21] That heat envelopes the spacecraft in a shroud of plasma that blocks communications with the ground. For a terrifying twelve to thirteen minutes, ground controllers can do nothing but wait for the shroud to clear and, if all goes well, receive a call from the astronauts verifying they are safe (a new antenna is being developed that will aid in reducing if not eliminating these reentry blackouts[22]).

As the ASTP Apollo capsule was beginning to be enveloped in the flames of reentry, the spacecraft experienced an electrical short in the communications system, and a high-pitched squeal filled the astronaut's ears. Stafford, who was reading items from the reentry checklist to his crewmates, had to yell as loudly as he could so that his commands could be heard.[23] At 80,000 feet, Stafford was supposed to make the call, "Inhibit RCS." The RCS, or reaction control system, is a series of nitrogen tetroxide thrusters that the spacecraft uses to control its attitude while in orbit and on reentry, basically keeping the capsule flying straight. "Inhibit RCS" instructs a crew member to flip a switch and turn the system off. Due to the squeal in their headsets and the crew yelling at each other to be heard over the noise, the order was either not given or not heard.

As the first parachute, what is called the drogue chute, deployed to slow the capsule's descent, the computers that control the thrusters sensed that the spacecraft was beginning to drift and sway back and forth in the thin upper atmosphere. The computer did what it was programmed to do and began firing the RCS—which should have been shut off—to try and stop the swaying.

As the spacecraft continued its descent and fell deeper into the atmosphere, vents opened, allowing the cabin to begin depressurizing and letting in fresh air. The problem was that the RCS was still firing and the fumes from the nitrogen tetroxide began filtering into the capsule, filling it with a poisonous yellow cloud.

"We got a sudden ingestion of fumes," Stafford later told reporters. "I had smelled nitrogen tetroxide gas in small amounts on my previous flights, so I knew what it was."[24]

Realizing what the cloud was, Stafford flipped a switch, cutting fuel to the thrusters, but by then, the astronauts—still in their spacesuits—were bathing in the vapors.

"We were coughing and wheezing," Stafford later said in an interview. "N204 is pretty bad stuff. It attacks the central nervous system."[25]

To add to the drama, the main parachutes did not automatically deploy and had to be deployed manually. When they finally unfurled, the spacecraft slowed but not enough to prevent it from slamming hard into the roiling Pacific Ocean, where it was immediately broadsided by a wave that flipped the craft upside down, leaving the astronauts hanging facedown from their seatbelts inside.

The fumes from the thrusters were subsiding but not enough to allow the crew to relax. Stafford ordered the crew to don oxygen masks, which were stowed behind the crew's seats. Stafford unbuckled his shoulder straps and, being upside down in the ocean, fell face-first into the nose of the capsule. Clawing his way back up to the couch, he pulled out the masks, handed them to his crewmates, and turned on the oxygen. Outside of the capsule, three airbags began to inflate and roll the spacecraft over. As the capsule slowly began flipping over in the ocean, Stafford looked up and saw that Brand's oxygen mask had slipped off his face and he was comatose.

Stafford raced to replace his crewmate's mask, and seconds later, Brand woke with a jolt and begins thrashing about. "He [Brand] hit me right here [in the jaw]," Stafford said, "and I went flying back. Then the mask came off [Brand] and he went out again. So this time, I fought my way over there . . . got a bear hug on him [and] got the mask on his face. He came to and started to fight but I wasn't about to let him go."[26]

Eventually the hatch was opened and the crew made it to the recovery ship. The follow-up medical exam and X-rays of their lungs showed that the astronauts had suffered chemical pneumonitis, a lung irritation caused by inhaling poisonous chemicals.[27]

The astronauts fully recovered from their terrifying ride, but the media had questions: specifically, who was at fault for not deactivating the RCS system. At a postflight press conference, Brand took full responsibility, but Stafford quickly stepped in, telling reporters that all three astronauts were to blame because each of them should have noticed that the switch wasn't flipped.

Aside from the near-disastrous landing, the Apollo-Soyuz mission was a huge success. The three astronauts and the director of the project, Glynn Lunney, were awarded Distinguished Service Medals from President Gerald Ford, who commented during the ceremony that the flight was a "great triumph of science and technology, an encouraging reminder in an atmosphere of mutual trust and respect that men from different countries, with different systems, can work together for a common goal with courage, intelligence, and success."[28]

8

GOOD MORNING TO OUR BEAUTIFUL WORLD AND TO ALL THE BEAUTIFUL PEOPLE WHO CALL IT HOME

In the indescribable blackness of space as you race across the heavens at five miles per second, the sea of shimmering stars, planets, and galaxies that glimmer above you slowly fade from view as a dark, black crescent appears. The crescent is shrouded by a thin layer of blue that slowly becomes thicker, its color getting brighter. An orange disk, the sun, peeks out over the crescent that is Earth, so bright that the stars above seem to run and hide. A brief rainbow of color bursts across the horizon.

Grabbing a camera, astronaut Anne McLain clicks off a shot of the spectacular light show going on outside one of the windows of the International Space Station then greets the people of Earth. "Good morning to our beautiful world and to all the beautiful people who call it home."

Soaring more than two hundred miles above Earth, every space farer will tell you that mornings in Earth orbit are spectacular. Of course, at five miles per second, an astronaut experiences sixteen sunrises in a single day, one every ninety minutes, but that doesn't mean that they experience "morning" that often. So how do astronauts and cosmonauts know what time it is?

It's simple enough when you're the only country flying in space. For example, from Project Mercury during the 1960s to the final flight of the space shuttle in 2011, the crews were flying on American central time. Things became more complicated when the International Space Station (ISS) began operations and there were flight controllers on two separate continents half a world apart. Operating on American central time put Russian controllers in a bind. The station teams generally operate on a twelve-hour day from 7:00 a.m.

to 7:00 p.m. in what amounts to a typical workday for mission controllers. Being seven hours ahead of their American counterparts meant that when it was quitting time at 7:00 p.m. Houston time, Russian controllers were calling it a day at 2:00 a.m., a long day to say the least. Unconfirmed stories suggest that the Russian team demanded a change in what time standard would be used because their local transportation stopped running at a certain hour.[1]

An agreement was reached between the two space superpowers that the space station and ground crews would operate on Universal Coordinated Time (the old Greenwich Mean Time) to make the schedule equitable for all parties.

Viewing the sunrise from ISS is always spectacular, but when it comes to an astronaut's morning routine, one fun tradition has been sadly missing from that routine—the morning wake-up call. For American astronauts, the wake-up call became a highlight of their morning routine when ground controllers began choosing songs with which to wake the crew. NASA's acting assistant administrator for congressional relations, Lynn W. Heninger, wrote that the purpose of the wake-up call was "to promote a sense of camaraderie and esprit de corps among the astronauts and ground support personnel."[2] Or as shuttle astronaut and STS-35 Capsule Communicator (CAPCOM) Kay Hire put it, the wake-up call "offers an opportunity for levity and a bit of shared camaraderie. It tends to stand out as a human element in an otherwise complex technical enterprise."[3]

The tradition of the wake-up call lasted through the final space shuttle flight in 2011 and has not been used during missions aboard the International Space Station. But how did this tradition begin in the first place, and will it ever return?

In 1963, the American space program began ramping up its quest to land a man on the moon as it moved from the one-man Mercury spaceflights to the two-man Gemini missions. Gemini was designed to be the lynchpin that connected our first tentative flights into space during Project Mercury with our ambitious goal of landing a man on the moon with Project Apollo. Each Gemini mission stayed in orbit for increasingly longer durations, incrementally testing and proving the technologies and techniques that eventually would land a man on the moon.

The fifth flight of the Gemini program, Gemini 6A, began its mission on December 15, 1965. As we have seen previously, the flight was renamed

"6A" after the aborted launch of Gemini 6 three days earlier, which nearly cost the space agency its booster and possibly the lives of the crew. The quick thinking of mission commander Wally Schirra saved the mission, and the crew was finally launched on December 15.

Gemini 6A was not a particularly long-duration flight, remaining in orbit for slightly more than twenty-four hours, but during that time, the tradition of the wake-up call was born. No one knows exactly why the tradition was started, but the theory is that it might have been a practical joke as NASA chief historian Bill Barry told Public Radio International.

"Wally was known as a jokester," Barry said. "I suspect he was surprised. I suppose there was some kind of inside joke, knowing Wally Schirra."[4]

As Schirra and his cabinmate, Tom Stafford, orbited Earth, a song began playing over the communications system. It was the voice of singer/songwriter Jack Jones singing a version of the title song from the hit Broadway and Hollywood musical, *Hello Dolly!* but this version was unique:

Hello, Wally.
This is Jack Jones, Wally.
It's so nice to know you're up where you belong.
All systems go, Wally.
You're 4-0, Wally.
Tom, all that's Navy Jazz for Razzamatazz
You can't go wrong.

While the Earth's turning,
The midnight oil was burning.
Gets you your requests from way back,
So, sit back with the wax, fellas.
Settle down and relax, fellas.
We'll see you down in Houston town again.[5]

From that moment on, wake-up calls became a part of American space folklore.

Eleven days before the launch of Gemini 6A, Gemini 7 was launched with astronauts Jim Lovell and Frank Borman onboard. Their mission was twofold—to set (at the time) a long-duration spaceflight record and to rendezvous in orbit with Gemini 6A and fly in close formation. It's true that the

Soviets had already attempted a similar rendezvous in 1963, launching two manned Soyuz spacecraft into orbit only a day apart, but the two spacecraft were only able to achieve radio communications with each other and never came closer than three miles of one another. Geminis 6A and 7 would fly within one foot of each other.

Following Gemini 6A's first wake-up call and during Gemini 7's thirteen-day mission, the crew was serenaded by many tunes ranging from operatic fare such as Chopin's *Les Sylphides*, Hungarian Rhapsody no. 2 by Franz Liszt, to "I Saw Mommy Kissing Santa Claus," a request by Lovell's twelve-year-old daughter Barbara, who hoped the song would bring her daddy home a little early for Christmas.[6] The latter selection added a new element to the musical tradition: songs would be selected by the capsule communicator (CAPCOM), the astronaut's family, or select friends.

Many times, the songs selected dealt with events scheduled for that particular flight day, such as a space walk or reentry. A favorite was the 1967 hit by Dean Martin, the lonesome and haunting, "Houston," with its chorus, "Goin' back to Houston, Houston, Houston," which was played on the last day of a mission as the astronauts prepared for their return to Earth.

The last test flight of the Apollo command module and lunar module, Apollo 10, looked hopefully to the future with Tony Bennett's, "The Best Is Yet to Come," an anticipatory song for the upcoming Apollo 11 flight and the first lunar landing.

During the Apollo missions, many of the songs paid homage to an astronaut's alma mater like "Hail Purdue" for Apollo 17's Gene Cernan or to the branch of service the astronaut served in. "Anchors Aweigh," for example, was played for Apollo 15 astronaut Jim Irwin.

When the Apollo program came to an end and the Skylab space station began operation in 1973, the three-man crew of Skylab 2 was greeted with the Herb Alpert and the Tijuana Brass song, "The Lonely Bull," and astronaut Pete Conrad (a navy captain) heard the shrill whistle of a boatswain's pipe call radioed aboard. Conrad appreciated the pipe's call but joked with mission control that they were a little late. "You should have started doing that on day two," he said, to which CAPCOM replied, "And the crew down here couldn't find a song for three bulls, it was only one."[7]

That was a problem these space-age DJs often faced during Skylab and the joint Apollo-Soyuz missions. NASA engineer Bob Parker said that during Apollo, "the wake-up music was usually picked the night before and that they sometimes had trouble finding a recording."[8]

Wake-up calls got a wake-up call of their own as the space shuttle took flight. With a larger number of astronauts coming from more diverse backgrounds and cultures, the wake-up call began to change. The biggest change was that the call not only included music, but also some comedy and special appearances by entertainers. Two of the captains from TV's *Star Trek* recorded special wake-up calls for the shuttle. On November 25, 1991, the shuttle crew aboard STS-44 woke up to the theme from *Star Trek: The Next Generation*, with Patrick Stewart performing the voice-over. After reading a short rewording of the series' introduction and a greeting, Stewart, in his Captain Jean-Luc Picard persona, gave the crew a hearty "make it so," Picard's iconic line from the series.

William Shatner, as James T. Kirk, the original captain of the *Enterprise*, also recorded a message for a shuttle mission. It was piped aboard the space shuttle *Discovery* on its final flight, and as one would expect, the greeting had a familiar ring to it:

Space, the final frontier. These have been the voyages of the space shuttle *Discovery*. Her thirty-year mission: To seek out new science. To build new outposts. To bring nations together on the final frontier. To boldly go, and do, what no spacecraft has done before.[9]

Creating a shuttle wake-up call became a badge of honor for entertainers, and they relished the chance to participate. Robin Williams reprised his role as armed forces DJ Adrian Cronauer from the movie *Good Morning, Vietnam*. On September 30, 1988, the crew of *Discovery* (STS-26) was jolted awake by Williams shouting Cronauer's famous, albeit modified, catchphrase, "Goooood morning, *Discovery*! Time to rise and shine, boys, and do that shuttle shuffle!" After a short bit of banter, Williams introduced a parody of the *Green Acres* television theme song, which was written by Houston disc jockey Mike Cahill, who thought previous wake-up calls were "kind of awful."

"So, what I did was thought, 'Gee, it would be nice to write, produce and custom design songs for the astronauts.' I picked three tunes that were pretty catchy that were very short and upbeat. *Green Acres* was the first one."[10]

Cahill thought the result was as "dopey and stupid" as he had hoped for:

On orbit is the place to be.
Freewheelin' on *Discovery*.
Earth rollin' by so far below,
Just give her the gas and look at this baby go!

We can't believe we made it here.
So high above the atmosphere.
We just adore that scenery,
Yeah, Houston's great but give me that zero G.

The combination of Williams's intro and the song was a huge, hilarious hit. Video from mission control shows engineers monitoring shuttle systems falling back in their swivel chairs, laughing, proving once and for all that NASA engineers do have a sense of humor.

The second flight of a space shuttle, STS-2, found another famous—or infamous—group of television space explorers making two calls to the space shuttle *Columbia*. On November 13, 1981, mission commander Joe Engle and pilot Richard Truly were awakened by the crew of the *USS Swinetrek* from *The Muppet Show*'s sketch "Pigs in Space," featuring the irrepressible Miss Piggy. As one would imagine, the puns were painful.

First Mate Piggy: It's time to contact the *Columbia* and speak to the astronauts.

Captain Hogthrob: Really and truly?

Dr. Strangepork: No, Engle and Truly.

The next morning, the *Swinetrek* crew greeted the astronauts once again with more moan-inducing puns, but years after the wake-up call sketch was relayed to *Columbia*, the joke would sound prophetic and tragic.

Captain Hogthrob: I hope they solved their problems with those heat tiles.

Dr. Strangepork: At least they've got tiles.

First Mate Piggy: Yeah, we've got shingles.[11]

Twenty-two years later, *Columbia* received its final wake-up call, "Scotland the Brave." The song was played for the crew's red team (to maintain twenty-four-hour operations, the shuttle crew was divided up into red and blue teams, which alternated shifts). Following the song, Captain Laurel Clark radioed back to mission control, "We are really excited to come back home." The next morning, *Columbia* broke apart during reentry, killing all seven astronauts. It was found that during launch, debris had damaged the shuttle's wing and tiles.

Once the International Space Station was manned and fully operational, the tradition of wake-up calls ended for the simple fact that during long-duration missions, the crew tends to wake up at slightly different hours. In 2020, however, the morning routine made a joyful return as a new era in spaceflight began. For the first time since the final shuttle launch in 2011, American astronauts rocketed into space from Cape Canaveral, Florida, aboard the SpaceX Crew Dragon capsule and spent sixty-two days aboard ISS. The morning of their return to Earth and splashdown in the Gulf of Mexico, the children of astronauts Doug Hurley and Bob Behnken sent a wake-up call of their own. Behnken's six-year-old son Theo was extremely excited to have his dad coming home. He was on a mission of his own.

"Wake up, wake up, wake up!" he shouted. "Wake up! Dad, wake up! Don't worry, you can sleep in tomorrow. Hurry home so we can go get my dog!"[12]

Behnken had promised his son a dog when he returned. For the record, at a postflight press conference, Behnken announced that Theo would be getting the puppy within two weeks once he understood how much work was involved in taking care of one. "Otherwise, it will be my dog instead of his."[13]

Wake-up calls also have made an unusual comeback in the least expected place—during unmanned missions to Mars. On July 4, 1997, Sojourner landed on the Martian surface, becoming the first of a long line of rovers to trek across the red sandy planet. One morning, mission engineers at the Jet Propulsion Laboratory (JPL) in Pasadena, California, decided to play a little joke and radioed a "wake-up" call to the rover. The song they transmitted to the little rover was another television theme song with an appropriate title.

Actor, writer, and comedian Paul Reiser is a man of many talents. Not only did he cocreate the Emmy Award–winning television series *Mad about You*, but he also wrote the show's theme song, "The Final Frontier."[14] On Sojourner's fourth day on Mars (or fourth Martian "sol"), the JPL engineers sent the rover a wake-up call reminiscent of the early days of manned space-flights. The song they chose was the *Mad about You* theme song.

The cast and production crews were surprised to learn that their song was the first ever played on Mars and quickly sent out press releases proudly boasting the news. They even used the video of controllers playing the song for the rover during the credits at the end of the show.[15]

In general, music has played an important role in mankind's exploration of space. Prior to the first manned flight into space, the Soviet Union launched an unmanned satellite and transmitted a recording of a song performed by a 110-member orchestra to the orbiting satellite, which in turn was relayed back to Earth. The Soviets' reasoning for this odd transmission was to "avoid arousing rumors that they had put a man into orbit."[16] Certainly they didn't orbit an entire orchestra.

In December 1965, the crew of Gemini 6A demonstrated their musical chops with a holiday prank. Only five hours after successfully rendezvousing and orbiting Earth only one foot away from Gemini 7, Tom Stafford radioed mission control: "We have an object, looks like a satellite going from north to south . . . very low, looks like he might be going to reenter soon. Stand by one. It looks like he's trying to signal us."[17]

Stafford and Schirra then broke out a set of bells and a harmonica that they had smuggled aboard the capsule and broke into a raucous rendition of "Jingle Bells," the first song ever played live from space.[18] NASA was not happy about the incident. Journalists from the *New York Times* reported that NASA officials "preferred to remain in the dark about how Captain Walter M. Schirra and Major Thomas P. Stafford got a harmonica and tinkling bells aboard Gemini 6." An unnamed NASA official was quoted as saying, "I'm sure it wasn't a case of smuggling."[19]

The flight of Gemini 6A might have been the first live music played from space, but it was far from the last. The most notable astronaut musician is Canadian Chris Hadfield. Hadfield flew on two space shuttle missions and is most remembered for his time aboard the International Space Station. As a member of the Expedition 34/35 crew, Hadfield broke out his guitar and, in a live televised event, sang a duet with Ed Robertson from the band Barenaked Ladies titled "I.S.S. (Is Somebody Singing?)." The song was cowritten by Hadfield and Robertson and was the first duet performed in space.[20]

Chris Hadfield performed live many times during his stay aboard ISS, including a rendition of Van Morrison's "Moondance" backed by the traditional Irish band the Chieftans in 2013. He later made headlines and became a YouTube sensation when he produced a video from the space station performing the David Bowie classic "Space Oddity."[21] The song is the story of a fictional astronaut, Major Tom, who is launched into space successfully but soon loses communications with ground control and becomes lost in space. The haunting video opens with ISS emerging from Earth's shadow. With the beautiful blue cloud-filled planet racing far below, the scene transitions to Hadfield floating from node to node singing the song. Hadfield was surprised by Bowie's reaction to the performance. "He described it as the most poignant version of the song ever done, which just floored me."[22]

When thinking about playing an instrument in outer space, like a flute, for example, one would think that it would be difficult to do, what with being weightless in a pressurized cabin, but as astronaut Katherine "Cady" Coleman proved, it's a breeze—at least for her.

As a member of the ISS Expedition 26/27 crew, Coleman brought her flute along with her and often found time to play in the station's cupola, a large six-window module used for viewing external station activities such as space walks and where the crew likes to sit and reflect as they watch spectacular views of Earth rolling far below.

"It's really different to play up here," she told reporters from the station in 2011. "I have been having the nicest time up in our cupola. I float around in there. A lot of times I play with my eyes closed."[23]

Unfortunately for Coleman, she was assigned to a mission in which none of her crewmates played an instrument. Back on Earth, Cady was part of a band that was composed of shuttle and space station veterans including Chris Hadfield. The band, called Bandella, plays an eclectic mix of bluegrass, jazz, folk-rock, and original compositions. During one of Coleman's personal moments aboard the station, they reunited the band via conference call. Bandella played from a conference room on Earth while Coleman played along on her flute aboard the space station.

As the band was preparing to call it a night, an alarm sounded on the space station. As trained, the crew literally dropped everything. For Coleman, that was her flute. Fortunately, it was a false alarm. When the all-clear was given, the crew went back to what they had been doing. Coleman decided to play her flute a while longer but realized the flute was missing.

"I had just let it go," she recalled. "Because it doesn't weigh anything in a way, you just don't have a sense of putting it down. We have a lot of different

During some free time, NASA astronaut and Expedition 27 flight engineer Cady Coleman plays a flute in the JAXA Kibo laboratory onboard the International Space Station. *NASA*

fan systems and air flow in the space station and after a while, you realize where things collect. And sure enough, I went to one of the most likely places . . . and there was the flute."[24]

Not only did Coleman bring her own flute with her on that mission, but she also brought a couple instruments from the Irish band the Chieftans: Paddy Maloney's pennywhistle and Matt Malloy's vintage flute. She also brought with her a flute owned by Ian Anderson, the founder of the rock band Jethro Tull.

In 1969 while Neil Armstrong and Buzz Aldrin were walking on the moon during the flight of Apollo 11, Ian Anderson and Jethro Tull were on a concert tour during which they played a song from their album *Stand Up* called "Bouree." The song was an arrangement of Bach's Suite in E Minor for Lute. Forty-two years later, Anderson teamed up with Coleman to do the world's first flute duet from space. With Anderson on the ground and Coleman aboard ISS, they played "Bouree" to commemorate the fiftieth anniversary of the first human spaceflight by Yuri Gagarin.[25]

Being an astronaut does have its perks, especially when it comes to scoring front-row seats at popular concerts. One of the most memorable space-age concerts was another space first. As rock star and former Beatle Paul McCartney was finishing a concert in Anaheim, California, on November 13, 2005, behind him a giant screen lit up with the images of astronauts Bill McArthur and Valery Tokarev live from 220 miles above Earth. McArthur playfully spun around doing zero-G backflips that amazed McCartney. "Before we go, can we see another one of those somersaults?"

After a short greeting between the crew and Sir Paul, McCartney broke into his song "English Tea" followed by an appropriate Beatles classic (and astronaut favorite), "Good Day Sunshine." McArthur wrapped up the live event by thanking McCartney, calling him an explorer in his own right, saying that astronauts and artists all target the same goal—ensuring a bright future for the young people of Earth.[26]

And that brings us back to where we started, hurtling around the world at five miles per second, experiencing sixteen sunrises a day. Good day sunshine, indeed. But there is one more astronaut wake-up tradition that is often overlooked, and it takes place even before launch—the time-honored steak-and-egg breakfast.

Like most traditions, this one began innocently enough. Alan Shepard was preparing for the first American manned spaceflight on May 5, 1961. It was a relaxed morning. He awoke at 1:10 a.m., showered, shaved, then traveled to Cape Canaveral's Hangar S, where he was joined by backup pilot John Glenn and the chief of the Astronaut Office, Deke Slayton. It's not clear who chose the menu for the country's first astronaut in space, but it was hearty—orange juice, a filet mignon wrapped in bacon, and scrambled eggs.[27]

From that moment on, astronauts from Mercury to Skylab began their journeys with the exact same breakfast. As the space shuttle rolled out of the hangar with a diverse crew flying each mission, the traditional preflight breakfast became more diverse as well. Steak and eggs didn't come off the menu but became somewhat archaic, pushed aside to make way for a more standard and lightweight meal such as cereal, oatmeal, toast, and juice.

At one time, NASA announced during preflight coverage of a mission what the astronauts had for breakfast, but that changed with shuttle flights as astronauts were given the freedom to decide whether or not they wanted to tell the world what they had for breakfast. Although there is conspiratorial chatter on social media sites claiming that this is a big NASA coverup, it is anything but. Some astronauts may have dietary restrictions that they want

The Apollo 13 crew enjoys the traditional steak-and-egg breakfast prior to suit-up and launch. From left: Fred Haise, Jim Lovell, Jack Swigert. *NASA*

to keep confidential, while others would just rather not get that personal. Some astronauts like to keep their Twitter and Facebook fans engaged during their mission, like Colonel Doug Hurley, who proudly posted a picture of—you guessed it—steak and eggs.

Although some in the shuttle and ISS astronaut corps might have chosen lighter, healthier morning meals, shuttle astronaut and commander of ISS Expedition 10 Leroy Chiao remembers reviving the tradition.

> I remember during one of my launch counts, the ladies were taking our prelaunch breakfast orders, going around the table. I was hearing things like dry toast. A little yogurt. Cereal. You gotta be kidding me, what kind of pantywaists am I flying with? They got to me and I replied firmly and evenly, "Steak and eggs, medium rare and over easy." Everyone looked at me funny. I stated the obvious. "Hey, we might go out tomorrow and get blown up. I'm going to have steak and eggs!" Immediately, three guys changed their orders to steak and eggs. I was doing all of us a favor, really. You need a hearty breakfast before launch. You're going to be really busy. Yogurt? Come on.[28]

9

HI, IT'S DOUG AND BOB
AND WE'RE IN THE OCEAN

It was the first successful launch of an astronaut from Cape Canaveral since the last launch of the space shuttle *Atlantis* in 2011. On May 30, 2020, Bob Behnken and Doug Hurley climbed aboard the ultra-sleek, science fiction–inspired Crew Dragon capsule *Endeavour* and rocketed skyward from historic Launch Pad 39A, where the Apollo moon missions once took flight more than fifty years prior. Two months later, after a successful docking with the International Space Station, the crew returned to Earth with a splashdown in the Gulf of Mexico just off the coast of Pensacola, Florida.

As the crew bobbed in the undulating Gulf waves patiently waiting for the recovery ship *Go Navigator* to haul the spacecraft aboard and free the duo from their capsule, Behnken and Hurley found that they had some time to kill. So what do you do? Make a few phone calls.

Behnken's wife, fellow NASA astronaut Megan McArthur, was watching the recovery with anticipation on television when her phone rang. Caller ID indicated that it was a spam risk, but instead of ignoring it, she decided to answer it anyway.

Also watching the recovery was flight director Anthony Vareha, who also received a call. Both McArthur and Vareha answered their phones and heard a familiar voice say, "Hi, it's Doug and Bob and we're in the ocean."

The pair could only chuckle. "Yeah," Vareha responded. "I can see that." According to the flight director, his capsule communicator (CAPCOM) Megan Levins said he should have answered, "Crap! Was splashdown supposed to be today?"[1]

During a postflight press conference, Doug Hurley took an introspective look back at the historic flight and the "prank" phone calls. "Five hours ago, we were in a spaceship bobbing around making prank satellite phone calls to whoever we could get a hold of, which was kind of fun, by the way. You can send him [SpaceX founder Elon Musk] the bill for the sat phone."

What may have been the first prank phone call made from a space capsule introduces us to a different side of the astronauts and engineers who are at the forefront of space travel and exploration, a lighter side that proves once and for all that astronauts and engineers are not completely buttoned up, rigid human beings. The general public usually only gets to see the professional side of these men and women—reserved, thoughtful, and disciplined. All true, but they are also human and come with all of the flaws, fears, and personalities of any other human being. They laugh, they love to play jokes on one another, they even—*gasp!*—curse.

Lighthearted anecdotes from the American space program could fill pages, with many of those stories coming from astronaut Wally Schirra. The Mercury, Gemini, and Apollo veteran was known as a true jokester and often regaled interviewers with his humorous behind-the-scenes tales, one of which he told Fran Foley during an interview for the Library of Congress's Veterans History Project. Foley asked Schirra about experiencing his first splashdown in 1962 aboard the Mercury *Sigma 7* capsule.

After successfully completing six orbits, the capsule's retro-rockets fired and Schirra made a textbook landing in the Atlantic Ocean only a half mile away from the recovery ship, the *USS Kearsarge*. Schirra jokingly radioed to mission control that he was so close that he thought they were "gonna put me on the number three elevator."[2]

The recovery plan called for a team of Navy Underwater Demolition Team (UDT) swimmers (now known as SEALs) to swim to the capsule and secure a floatation collar around it to prevent it from sinking with the astronaut aboard. This was in response to the near-tragic incident during the second Mercury flight in which the hatch of Gus Grissom's *Liberty Bell 7* capsule was blown off, causing it to fill with water and sink to the ocean floor. The astronaut, with water filling his spacesuit, helplessly waited for help to arrive.

With *Sigma 7* rolling in the ocean's waves, the swimmers removed panels from the top of the capsule to let fresh air in and through which Schirra could hear the swimmers outside.

"I hear this unbelievable splashing, yelling and screaming," Schirra recalled. "All of a sudden everything lurches and this . . . guy in the water leaps up on the top [of the capsule] and I said, 'What in the hell is going on out there?'"

The swimmer, who is trained to have nerves of steel and to perform some of the most dangerous and terrifying covert missions, shouts back, "I saw the biggest jellyfish of my life here in the Pacific. It's unbelievable."

"What color was it?" Schirra asks.

"Orange and white!" the swimmer replies.

Schirra laughs, "You never saw a parachute under water before?"[3]

Schirra would fly the fourth mission of the Gemini program, Gemini 6A, in December 1965, when the first space practical joke was pulled. During the flight, his crewmate Tom Stafford surprised mission control when the crew reported sighting a strange object. "We have an object, looks like a satellite going from north to south up in a polar orbit. He's in a very low trajectory traveling north to south and has a very fineness ratio, looks like it might even be a ball of sticks."[4]

In a later interview with *Smithsonian Magazine*, Stafford said he could hear the voices of mission control becoming tense[5]; after all, the astronauts had spotted something unusual and potentially dangerous in orbit.

Stafford continued his report to ground controllers, "It's very low, looks like he might be going to reenter soon. Stand by one. It looks like he's trying to signal us."

The report ended with a rendition of "Jingle Bells" played on harmonica accompanied by a set of bells. There was a pause before the crew of Gemini 7, who had just completed a rendezvous with Stafford and Schirra chuckled over the radio, "We see them, too!" CAPCOM also got a laugh. "You're too much, six."

Stafford said the prank was weeks in the making. "Wally came up with the idea. He could play harmonica, and we practiced two or three times before we took off, but of course we didn't tell the guys on the ground. We never considered singing since I couldn't carry a tune in a bushel basket."[6]

Schirra's final mission occurred in 1968 with the flight of Apollo 7. This would be the first flight of the redesigned command module, which would eventually send men to the moon but had claimed the lives of Virgil Ivan

"Gus" Grissom, Edward White II, and Roger Chafee during a prelaunch test on the launch pad at Cape Kennedy only a year before.

One of a myriad of skills required for a successful flight to the moon is the ability to navigate by the stars and constellations. Even though Apollo 7 would be an Earth orbital mission, the crew still was required to learn the stars, planets, and constellations. It was during this training that the crew of Apollo 7 learned that their fallen companions from Apollo 1 had left behind a little joke.

Schirra and his crewmates Walt Cunningham and Donn Eisele flew to Morehead Planetarium in Chapel Hill, North Carolina, where they met with Tony Jenzano, who oversaw the celestial training of the Apollo astronauts. Jenzano was an expert in the field of celestial navigation and taught the astronauts every possible way of identifying stars, orienting with constellations, and calculating checkpoints that would be used to pinpoint their spacecraft's exact position on its journey to the moon. One of the Jenzano's methods for training the men was to seat the astronauts in chairs with a reproduction of their Apollo capsule's window in front of them. Star fields were projected onto a wall in front of the windows to represent what the patterns would look like in space.

The astronauts continued their training at the Griffith Planetarium in Los Angeles, where they were greeted by the planetarium's director, Dr. Clarence Cleminshaw, who told them that there were three new stars in the celestial system that had been only recently discovered. Cleminshaw said that he learned of these stars from the Apollo 1 crew and Tony Jenzano. The stars were named Navi, Regor, and Dnoces. The Apollo 7 crew began laughing when they realized that the names of the stars were actually the names of the Apollo 1 crew spelled backwards—"Ivan" (Gus Grissom's middle name), Roger, and "Second" (Ed White II).[7] For at least two years, the director of the planetarium taught the Apollo astronauts those names, never once catching on to the joke.

During the early days of manned spaceflight, an astronaut never knew when he would be pranked by a fellow astronaut. Take the flight of Apollo 12, for example. The three-man crew of Commander Charles "Pete" Conrad, command module pilot Richard Gordon, and lunar module pilot Alan Bean would be the second manned landing on the moon. Their mission was to not only explore the lunar surface, deploy experiments, and gather moon rocks, but also to bring back pieces of the robotic Surveyor 3 spacecraft that had landed on the moon only two years before.

To keep track of the enormous number of tasks the astronauts had to perform, NASA provided checklists. The checklists were kept in spiral-bound notebooks attached to the astronauts' wrists and would guide them step-by-step through their activities. Each page of these cuff checklists was three-and-a-half inches square and documented every move the astronauts would make both in the command module and on the moon. Everything was detailed within these pages—"Walk to Camera Area" meant to adjust the camera angle, "Exit Feet First" reminded them how to egress from the lunar module, and so on.

As the astronauts went about the tasks at hand, kangaroo-hopping around the lunar surface, they noticed that cartoon drawings of a Snoopy-like dog had been scribbled on the pages of the checklists with captions like "They have doggie bags on the moon?" In the instructions for reentering the lunar module, the dog says, "Well, back to my beloved tennis shoes." One cartoon showed the dog stretching a cable from one of the lunar experiments to a dynamite plunger.

The drawings were made secretly by NASA preflight operations chief Ernie Reyes as a joke on the unsuspecting crew. Snoopy was used in the cartoons to pay homage to the Charles Schulz cartoon character and the black-and-white communications caps the astronauts wore that were affec-tionately called "Snoopy caps."

Another prank really took the crew by surprise. It was about two and a half hours into the second extravehicular activity (EVA) on the moon's surface when Bean turned the page of the checklist and was greeted by the smiling face of a topless Cynthia Myers, *Playboy* magazine's Miss December 1969. The caption scribbled in a speech bubble emanating from the model's mouth read, "Don't forget—Describe the protuberances."[8]

Bean hopped over to Conrad to show him what he had found. Conrad lifted his wrist to show that he, too, had a photo—Miss September 1967, Angela Dorian.

"We didn't say anything on the air," Bean said in an interview with the magazine in 1994. "We thought that some people back on Earth would

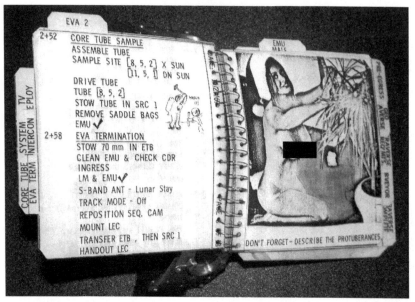

The Apollo 12 crew had quite the surprise when they opened their cuff checklist while exploring the moon in 1969—a little holiday cheer from *Playboy* magazine's Miss December. *NASA*

become upset if they found out we had Playboy Playmates in our checklists. They would have said, 'This is where our tax money is going?'"[9]

"We giggled and laughed so much," Conrad confessed, "that people accused us of being drunk or having 'space rapture.'"

Years later, after the *Playboy* incident had faded to memory, Al Bean looked at a photo he had hanging on a wall. It was a photo of himself that had been taken by Pete Conrad on the moon. Looking closer, he noticed that his cuff checklist was open. "Holy Christmas!" he said. "That's the Playmate of the Month sitting on my arm!"[10]

In the photo, his wrist notebook was wide open to the page with the Playboy Playmate on it for all the world to see. Though Bean and Conrad tried to keep the joke a secret, Miss September had unintentionally photobombed the image.

A classic practical joke was not pranked on the astronauts, but the engineers manning launch control consoles at Cape Canaveral just prior to John Glenn's orbital flight aboard Project Mercury's *Friendship 7*. According to NASA engineer Manfred "Dutch" von Ehrenfried, the days leading up to Glenn's launch were fraught with troubles and the launch had to be scrubbed several times for various reasons. In the interim, NASA's first flight director, Chris Kraft, kept the team on their toes with long hours of simulation training sessions and meetings to review flight plans and rules.

One of Kraft's reliable directors, a young Gene Kranz, thought he would break up the monotony and began work on an elaborate practical joke. Kranz teamed up with video coordinator John Hatcher and cooked up a scheme that had Kraft fuming.

It was another day of repeated launch simulations. As the unfueled Atlas rocket sat quietly on the launch pad, engineers in mission control performed status checks, flipped banks of switches, and worked the countdown all the way to zero. As always, the rocket stayed firmly latched to the launch pad. It was unfueled. This was only a test, after all.

During one simulation, Kraft strolled up to Kranz and asked how it was going. Without looking up or cracking a smile, Kranz reported that everything was going smoothly. The countdown continued: three, two, one—*zero!* Kranz flipped a switch and at that same moment, Hatcher flipped a switch himself. On the video monitor, Kraft watched as the engines of the Atlas rocket burst into flames and the shining silver missile began to rise from the ground.

"Did you see that?" Kraft shouted excitedly.

"See what?" Kranz replied nonchalantly.

"The damn thing lifted off!" Kraft yelled.[11]

Hatcher had patched in a video of a previous Atlas rocket lifting off on the video monitor. Glenn's rocket still sat silently on the pad. Kraft was furious and demanded, "Who the hell did this?"[12]

He never got a reply, only laughter from his team. Realizing that he had been had, Kraft could only shake his head and chuckle.

Even the best-laid plans for a prank can land with a thud or even worse. A case in point happened in 1991 during the flight of the space shuttle *Atlantis* during the STS-44 mission. It was during the early morning hours of Thanksgiving Day. The crew of *Atlantis* was sound asleep as the spacecraft passed over South Africa, an area where there were no communications links and the shuttle was out of contact with mission control.

The stillness of the flight was broken when lead flight director Milt Heflin found himself hurled into a crisis. The Air Force tracking station that monitors space debris, Cheyenne Mountain in Colorado Springs, relayed a warning to mission control in Houston that a disabled Turkish satellite was heading straight for the shuttle. Normally, "the Mountain," as the tracking station was known, provides twenty-four-hour notice of approaching debris to the flight director. That is more than enough time for the shuttle crew to perform evasive maneuvers. This time, the shuttle had only fifteen minutes.

Heflin's associates have had nothing but glowing comments about him and recognize him as one of the best flight directors of the shuttle program. Even during the most difficult and dangerous situations, Heflin remained cool and calm and rarely, if ever, became rattled. "When I think about all my time [as flight director]," Heflin told the online website ARS Technica, "I don't remember being so nervous or upset about something as I was then."[13]

Knowing that the situation was completely out of his control and the shuttle was out of communications with the ground, Heflin left his position to regain his composure and think about what could be done to avert disaster. On the way out of the control room, he was stopped by one of the flight controllers who tried to direct his attention to a console that tracked the position of the shuttle as it orbited Earth with an animated display. It also would track the incoming Turkish satellite.

"I don't need to see the damn thing!" Heflin barked.[14]

When Heflin returned to the control room, he was met again by the controller. This time, Heflin listened to what the engineer had to say. On the tracking display, he saw the "Turkish" satellite—a plump, round, animated Thanksgiving turkey proudly waddling in orbit around the Earth. Someone had programmed this animated butterball into the display software.

Heflin was not amused and told the controller that he would speak to him after his shift had ended. At that moment, a representative from the Shuttle Program Office came in and asked about the object that threatened *Atlantis*. Heflin explained the situation and the representative turned white as a ghost. The staff at the shuttle office had heard about the peril *Atlantis* was in and immediately awakened the deputy manager of the shuttle program, Brewster Shaw, to alert him to the impending disaster. The joke was quickly running up the chain of command.

When his shift had ended, Heflin stayed awake the rest of the night typing a letter to Shaw explaining what had happened. When Shaw arrived at mission control the following morning, Heflin handed him the letter and began telling the manager about the prank. "As I tell Brewster this story," Heflin recalls, "[Brewster] is having a hard time not laughing."[15]

The flight director promised that he would take care of the situation, and Shaw let the matter go.

Flight simulations are never boring for the engineers who are entrusted with protecting the astronauts' lives. Any number of glitches are programmed into the training to ensure that the teams are at peak readiness when the actual spaceflight begins. In the case of flight surgeons, it's a bit different. They are required to be on station at their consoles during a simulation "just in case," but many times, there just isn't much for them to do and the hours of waiting can become boring.

Flight controller Joseph DeAtkine realized this and one day decided to spice things up for the flight surgeons on duty during a launch simulation. DeAtkine's team went to a local Houston hospital and obtained an EKG of an actual patient who had a heart attack and died.

As the countdown clock ticked away during the simulation, DeAtkine imported the EKG data into the simulation software. When the dress rehearsal finally ended a few hours later, the team met to assess their performance.

"The training people came online and [told us] whether you did [this] right or did this wrong or you missed that," DeAtkine said. "But when it came to the flight surgeon, they said, 'Oh, by the way, astronaut so-and-so died.' Of course, all of Mission Control just busted out laughing."[16] The flight surgeon completely missed the EKG reading.

After the cancellation of the Apollo program, NASA had some leftover hardware including a few Saturn V and Saturn IB boosters and a couple command and service modules. They put these leftovers to good use with the development of the agency's first space station, Skylab.

During the final mission of the space station Skylab 3, astronaut Robert Crippen was manning the communications console in Houston, keeping close contact with astronauts Owen Garriott, Alan Bean, and Jack Lousma, who were orbiting Earth and setting what would be, at the time, a space duration record.

Controllers were shocked when a female voice suddenly broke into the communications. "Hello, Houston?" the feminine voice said. "This is Skylab. Are you reading me down there?"

The controllers were baffled. At this time, NASA had not yet flown a woman in space. It was an all-male crew aboard the space station, but the voice clearly identified itself as calling from Skylab. Crippen at the CAPCOM console looked shocked. "Well, hello, Skylab," he said. "Who is this?"

"Hi, Bob. This is Helen," the voice replied.

Helen? Owen Garriott's wife? Aboard Skylab?

"What are you doing up there?" Crippen asked.

"We just came up to bring the boys a fresh meal, a hot cooked meal," the voice said. "They haven't had one for quite a while. We thought they might enjoy that."

Crippen continued questioning the voice. "How did you get up there?"

"Oh, we just flew up," she replied. "We've been looking at forest fires that they have all over California. It's a beautiful sight from up here."

The other controllers in the room gathered around Crippen's console. *No, it couldn't be Helen Garriott aboard Skylab. Could it?*

"Well, I see the boys are floating in my direction," she concluded. "I've got to get off the line. I'm not supposed to be talking to you. See you later, Bob."

And with that, communications with the voice ended. Flight director Neil Hutchinson looked at Crippen and said, "Bob, what's going on?"

Crippen replied. "You heard it the same time I did. I don't know. . . . Well, obviously, she wasn't there. I don't think she was."[17]

The CAPCOM couldn't hold out any longer and started to laugh. As Owen Garriott explained in an interview years later, it was an elaborate and well-planned practical joke.

> The story started two or three months before I flew. At the time I made a recording of my then-wife Helen as if she were visiting the Skylab . . . to bring us a home-cooked meal. Bob Crippen was going to be on [as CAPCOM] maybe twenty minutes before I was going to be passing over a ground station. . . . [I radioed] that I would have something special for you on the next trip. Bob knew exactly what I was talking about. So twenty minutes later when we came back in contact with the ground, a female voice came on the radio channel. It was a proper radio channel so they knew it wasn't something done locally. It was actually coming down from space. . . . Bob Crippen . . . knew what was going to happen ahead of time. He had that discussion already pre-planned on a sheet of paper. They never figured out how this happened. . . . I told them after twenty years how we did this trick.[18]

These anecdotes are proof that engineers and astronauts are human, and just like anyone else, frustrations when things don't go quite as planned can cause that person, even an astronaut, to blurt out a curse word or two. And if you are a military test pilot, as America's first astronauts were, those words are just part of your everyday vocabulary and can slip out even during the most guarded conversations. However, for astronauts, there is always the chance of letting it fly on a hot microphone broadcasting to a worldwide audience.

One could only imagine that the Mercury astronauts, being military test pilots, might let a curse word or two slip out during flights, but that wasn't the case, although they once came close. In 1962 an odd question was posed by the chief of the Astronaut Office, Donald "Deke" Slayton, to one of the Mercury astronauts just seconds after liftoff, but the astronaut onboard, Wally Schirra, was too quick for him.

Once again, it was during Schirra's flight on *Sigma* 7, the fifth manned Mercury flight. On October 3, 1962, after a short delay due to issues at a radar tracking station, the Atlas rocket's engines finally ignited and the capsule carrying Schirra slowly climbed away from Earth. With the world watching, the silver cylinder gained speed and coursed skyward. Three minutes into the flight, Slayton radioed to Schirra and asked, "Hey, Wally. Are you a turtle?"

As the story goes, that question is used to identify members of the Ancient and Honorable Order of the Turtles, a fictitious group organized by Captain Hugh P. McGowan of the US Army Air Corps during World War II as a bit of relief from the horrors and dangers of war.[19] The group poked fun at the upper-class social organizations of Europe but, unlike those snooty societies, the Order of the Turtles had no constitution and required no dues of members nor social status. It was just a way to have some fun. Like many societies, there was a secret greeting for members: are you a turtle? The correct answer is, "You bet your sweet ass I am." Not answering correctly meant that person would be buying the questioner a drink.

Schirra, on worldwide radio and television, was asked if he was a turtle. Not wanting to curse on the air but definitely not wanting to buy Slayton a drink, the quick-thinking astronaut replied, "Roger," and momentarily switched over to the capsule's voice recorder.

During his postflight debriefing aboard the recovery ship *USS Kearsarge*, one of the men in attendance demanded to know what the astronaut's response was. Schirra flipped on the recorder for all gathered around to hear:

Slayton: Hey Wally, are you a turtle?

Schirra: You bet your sweet ass I am.[20]

Following Gordon Cooper's twenty-two-orbit Mercury mission one year later, the Mercury 7 astronauts found themselves being greeted by President Kennedy at the White House. Schirra was taken aback when the president asked, "By the way, Wally, are you a turtle?"

"I had to think twice," Schirra said. "[You couldn't say] 'you bet your sweet ass' to the president of the United States."[21]

There is no indication that Schirra bought the president a drink.

While Wally Schirra was conscious of his language when on a mission, others weren't as disciplined. Such words were just part of their vocabulary. Take the flight of Apollo 10, for example. The mission would be the dress rehearsal for the first manned landing on the moon. Astronauts Tom Stafford and Gene Cernan would test the lunar module in orbit around the moon, flying within nine miles of the surface before redocking with John Young, who was orbiting nearby in the command module and returning to Earth.

During the flight, the air force and navy pilots regularly slipped profanities into their conversations, causing some consternation back in Houston and in the Nixon White House.

After the Saturn rocket's third-stage engine ignited and sent the astronauts out of Earth orbit in a maneuver called translunar insertion (TLI), one of the crew exclaimed, "Man, that TLI is quite a ride! Never forget that son of a bitch."[22] While discussing navigating by stars, command module pilot John Young commented on what he saw: "I can see stars in the day. I can see something in the day. They're not stars. Shit, it's a bunch of crap."[23]

Later on as he fumbled with a star chart, Stafford let his feelings be known about the situation saying, "How come these bastards gave us this star chart [laughs] with no Velcro?"[24]

And as the lunar module flew above the lunar landscape, Stafford spotted the crater Censorinus, with impressive shadows highlighting the giant boulders within. "I've got Censorinus here! It's bigger than shit!"[25]

A reporter in the media room at mission control overheard the comment. He turned to astronaut Jack Schmitt, who was also in the room, and asked, "What did Colonel Stafford just say?"

Schmitt replied, "He said, 'oh, there's Censorinus. [It's] bigger than Schmitt.'"

In all, the crew of Apollo 10 cursed 230 times, enough to catch the ear of the president of the nondenominational Miami Bible College, Dr. Larry Poland, who was listening to the mission back on Earth. Poland was one of many people who registered complaints with the White House about the language. In an interview with the *Orlando Sentinel*, Poland said, "I've gotten calls from many people who were astounded that they were broadcasting things like that 240,000 miles from [Earth] when it's the kind of language you would expect to see on the restroom wall."[26]

The complaints fell on deaf ears, however, and the astronauts of Apollo 10 successfully completed their mission, cursing and all.

That wasn't the last time Apollo astronauts would let a few choice words fly but they were nothing compared to what happened on the flight of Apollo 16, when the mother of all four-letter words spilled out.

Following a record-breaking seventy-one hours on the moon, during which time the astronauts of Apollo 16 had ventured outside the relative safety of their home away from home, the lunar module (LM), to drive the lunar rover 16.6 miles and collect 209 pounds of lunar rock and soil, astronauts John Young and Charlie Duke returned to the small, cramped quarters of the LM and prepared the craft for liftoff to rejoin their crewmate, Tom Mattingly, who had been orbiting Earth's barren satellite in the command module.

As they prepared for lunch (a little turkey and gravy), the astronauts began a debriefing call with mission control. The official NASA transcript of the conversation read like this:

Mission Control: You guys did an outstanding job.

Young: I got the gas again. I got them again, Charlie. I don't know what gives them to me. I think it's acid stomach I really do.

Duke: It probably is.

Young: (Laughing) I mean, I haven't eaten this much citrus fruit in twenty years! And I'll tell you one thing, in another twelve days, I ain't never eating any more.

Well, that's not how the conversation *actually* went, but that was the readout from the "official" transcript. Eventually, the actual audio recording and a revised transcript was released by NASA, and it became quite clear that the original printed readout had been scrubbed for public consumption when Young talked about his "gas" and then slipped in "that" word:

Mission Control: You guys did an outstanding job.

Young: I got the farts again. I got them again, Charlie. I don't know what gives them to me. I think it's acid stomach I really do.

Duke: It probably is.

Young: (Laughing) I mean, I haven't eaten this much citrus fruit in twenty years! And I'll tell you one thing, in another twelve fucking days, I ain't never eating any more.[27]

It didn't end there. A few short minutes later, Young misplaces something in the lunar module.

Young: What did I do with them?

Duke: What did you do with them?

Young: They're right over the . . . Well, they're gone. I put them . . . right up in here. They ain't there? Oh, shit. Must be on the floor.

At this point, Houston interrupts the conversation.

Mission Control: Orion, Houston.

Young: Yes, sir.

Mission Control: Okay, John. We have a hot mike. . . .

Duke: Sorry about that.[28]

There is another side of space exploration that leaves us shaking our heads and chuckling. It doesn't involve the astronauts, engineers, or NASA. Let's just call it "space-ploitation." As it was back in the day when traveling salesmen roamed from town to town selling their "magic elixirs" or deeds to nonexistent gold mines, modern-day con artists still exist today, taking advantage of those willing to believe whatever bill of goods they are trying to sell.

In 1955, five-year-old Nancy Munie was given a unique gift by her uncle, H. W. "Dick" Miller—a deed to a plot of land, but not just any piece of land. The deed was for beach rights to the Sea of Tranquility on the moon.

The official-looking document, complete with fancy gold trim and a brochure detailing the property, was purchased from the Interplanetary Development Corporation and quickly became a family joke. According to Munie, the document was used as a warning against bad behavior. "My mom would say, 'watch out because I'm going to send you to that acre of land on the moon.'"[29]

Nancy was attending college at Eastern University in Charleston, Illinois, in 1969 when Neil Armstrong took mankind's first step onto the moon. A local newspaper caught wind that the young woman had this deed, and soon after, the St. Clair County probate office extended an invitation to officially register the deed. Cameramen were in attendance to record the moment as the probate clerk accepted the $3 filing fee and registered the lunar deed.

Three months later, Munie's grandfather, David Mantle, wrote to NASA:

It appears that Apollo 11 landed on my acre of the moon. Consequently, I would greatly appreciate being furnished with a 145th sample of rock and soil collected by the astronauts from my moon holdings.[30]

Her grandfather signed Nancy's name to the letter and sent it off. Remarkably, NASA Associate General Counsel E. M. Shafer replied:

You appear to be aware already that your deed is an interesting but efficacious document. Legally, the grantor conveyed to you only the same right, title and interest in the area described in the deed as he himself possessed. Unfortunately for you, he possessed absolutely nothing.[31]

As a sort of consolation prize, Shafer attached four ten-cent "Man on the Moon" stamps and a four-cent Project Mercury stamp to his reply. Munie went on to finish college and start a career and a family. When asked about the deed during Apollo 11's fiftieth anniversary in 2019, she replied, "When they started talking about what happened fifty years ago, it kind of dawned on me how old I am."[32]

Nancy Munie's story is an endearing tale of an uncle gifting a unique heirloom to his niece, which made for many family memories. In that regard,

Nancy and her family were richer for it. However, many others are regularly fleeced by con artists selling pieces of the moon. In fact, the moon has been "sold" for as long as anyone can remember. One man, Dennis Hope of the so-called Lunar Embassy, continues to sell chunks of Earth's lifeless satellite today at $20 a pop. Hope claims he can sell the moon because, despite the 1967 Outer Space Treaty that forbids it, he has declared himself the owner of several parcels of land across the universe. He says that he has sold more than 611 million acres of the moon and 325 million acres of Mars.[33]

As the Canadian con man and riverboat gambler Canada Bill Jones was quoted in the late 1800s, "A sucker has no business having money in the first place."

10

"JUST IN CASE" IS THE CURSE OF PACKING

How many times have you heard that immortal packing phrase, "just in case"? Probably too many times to count. You have probably uttered those words yourself at one time or another when preparing for vacation travel. Leaving on a trip to Hawaii? "I'll pack this goose-down parka . . . *just in case*." Spending a day on the beach? "I'll take along all five *Game of Thrones* books to read . . . *just in case*."

OK, maybe that's a little over the top, but face it, we all have trouble deciding what to bring with us when going on a trip. It's that age-old traveling conundrum: you don't want to bring everything you own on your trip but there's always that nagging fear that you'll get caught off guard and utter those equally famous travelling words, "I should have brought . . ." As author Alexandra Potter once wrote, "'Just in Case' is the curse of packing." Now, imagine if you had to make that decision but could take only items with a total weight of 1.5 pounds and the nearest Walmart, Target, or Walgreens is more than a quarter of a million miles away?

As human beings broke the chains of Earth's gravity and began their quest to conquer space, it quickly became obvious that the men and women flying into the final frontier would need something to make themselves feel more comfortable and at ease during their perilous journey, a little piece of home to provide them with some sort of physical connection with the Earth, their families, and their friends. Enter the official flight kit (OFK) and personal preference kit (PPK).

In 1938, the *Code of Federal Regulations* (CFR) was established to codify all of the federal government's rules and regulations into one neat document.[1] Needless to say, the document has grown exponentially over the years, with new sections being added to meet the ever-changing world and technologies around us. Within the section titled the *Electronic Code of Federal Regulations* (eCFR), Title 14 was established to govern rules and regulations for aeronautics and space research. Since the days of the two-man Gemini missions in the mid-1960s, section 1214.600 of this code has detailed what PPK and OFK are and what they may contain. The following are entries from the code during the space shuttle era:

> 1214.601 (c) *Personal Preference Kit (PPK).* A container, approximately 12.82 centimeters × 20.51 centimeters × 5.13 centimeters (5"×8"×2") in size, separately assigned to each individual accompanying a space shuttle flight for carrying personal mementos during the flight.[2]

> 1214.602 (a) *Premise.* Mementos [flags, patches, minor graphics, medallions, insignia and similar small items of little commercial value especially suited for display by the individuals or groups to whom they have been presented] are welcome aboard Space Shuttle flights. However, they are flown as a courtesy, not an entitlement.[3]

The document further outlined the procedures astronauts had to follow to seek approval for items in their PPK:

> At least 60 days prior to the scheduled launch of a particular mission, each person assigned to the flight who desires to carry items in a PPK must submit a proposed list of items and their recipients to the Associate Director, NASA Johnson Space Center. The Associate Director will review the proposed list of items and, if approved, submit the crew members' PPK lists through supervisory channels to the Associate Administrator for Human Exploration and Operations for approval. A signed copy of approval from the Associate Administrator for Human Exploration and Operations will be returned to the Director, NASA Johnson Space Center, for distribution.[4]

That sounds like a lot of red tape to navigate in order to take a photo of your family with you on your journey, and it is, but for good reason. There are limits to how much weight a booster rocket can hurl into orbit or into the solar system. Weight also plays a critical role in the reentry process, so every extra ounce has to be accounted for.

The OFK differed from the PPK in that it enabled "NASA, developers of NASA sponsored payloads, NASA's external payload customers, other federal agencies, researchers, aerospace contractors, and counterpart institutions of friendly foreign countries to utilize mementos as awards and commendations or to preserve them in museums or archives."[5]

During the Mercury missions of the early 1960s, astronauts mostly adhered to the then less restrictive and unwritten law. For the most part. Once again, we begin with astronaut Wally Schirra and the flight of *Sigma* 7 in October 1962, only this time, Schirra would be the one pranked.

Astronaut Gordon Cooper, who would fly the last Mercury mission aboard *Faith* 7 in 1963, and Indy race car driver Jim Rathman picked up an airplane-size bottle of Cutty Sark scotch and a pack of Tareyton cigarettes with every intention of smuggling them aboard *Sigma* 7. In keeping with weight limitations, Cooper stripped the pack of cigarettes down to only four then secretly stashed the alcohol and cigarettes in a small compartment of the spacecraft's instrument panel.

Schirra's flight was going smoothly as he orbited Earth and tested the spacecraft's systems and controls. Schirra looked behind a panel with a red flag on it that read, "remove before flight" and discovered the hidden contraband. Schirra got a laugh out of it but didn't partake in the alcohol. He did, however, get the last laugh.

"I drank the scotch as soon as I had a chance on board the recovery carrier," he said in an interview years later. "The medics all wondered why I had a small alcohol level in my postflight blood tests! I guess a nicotine level would have really thrown them."[6]

The most poignant and to many the most famous moment in spaceflight history was the historic flight of Apollo 8, the first time that humans ventured away from our planet to orbit another celestial body and experience earthrise. Commander Frank Borman, Jim Lovell, and William Anders entered lunar orbit on Christmas Eve 1968, during which time they read from the Book of Genesis to an estimated one billion viewers back home on Earth.

The crew concluded the broadcast with the words, "good night, good luck, a Merry Christmas, and God bless all of you, all of you on the good Earth." When the telecast ended, the three men prepared for Christmas dinner—turkey with gravy, cranberry sauce, grape juice, and a little surprise. The chief of the Astronaut Office, Deke Slayton, had smuggled aboard three two-ounce bottles of 100-proof Coronet VSQ California grape brandy. Upon finding the booze, mission commander Frank Borman became incensed.

"When we opened up the dinner for Christmas," Borman explained in a 1999 oral history interview for the Johnson Space Center, "and I found somebody had included brandy in there, you know, I didn't think that was funny at all. Because you and I both know, if we'd have drunk one drop of that damn brandy and the thing would have blown up on the way home, they'd have blamed the brandy on it. You know, I wanted to do the mission and I didn't care about the other crap."[7]

Borman's crewmates were in complete agreement and the bottles returned to Earth unopened. Lovell eventually auctioned off his bottle for $17,925.[8]

The first time an American astronaut was allowed to bring alcohol into space and consume it was on Apollo 11 during the first landing on the moon. If Neil Armstrong and Buzz Aldrin landed successfully, the lunar module pilot, Aldrin, wanted to conduct a religious service of some sort.

Months before the launch, Aldrin approached Deke Slayton with the idea and to get his approval. It wasn't that Slayton was against holding such a ceremony—after all, it would be mankind's greatest achievement—but at that time, NASA and the agency's director Thomas Paine were embroiled in a lawsuit brought against them by devout atheist Madalyn Murray O'Hair. She contended that reading from the Book of Genesis during the flight of Apollo 8 violated the Constitution's First Amendment and the separation of church and state.[9] Slayton granted approval for Aldrin's service but advised the astronaut that millions of people from around the world with wide-ranging religious beliefs would be watching, and he should keep his comments less controversial.

Aldrin agreed and kept his word. As soon as Neil Armstrong uttered those now-immortal words, "the *Eagle* has landed," Aldrin began to hold communion on the moon, but first, he made a brief statement to the world. "I'd like to take this opportunity to ask every person listening in, whoever and wherever they may be, to pause for a moment and contemplate the events of the past few hours and to give thanks in his or her own way."[10]

Off-mike, Aldrin took out a golden chalice that he had been given by the minister of his church, Reverend Dean Woodruff, a communion wafer, and a small vial of wine and began the service.

"I poured the wine into the chalice our church had given me," Aldrin later recalled. "In the one-sixth gravity of the moon, the wine curled slowly and gracefully up the side of the cup. It was interesting to think that the very first liquid ever poured on the moon, and the first food eaten there, were communion elements."[11]

The chalice with which Aldrin performed communion at the Sea of Tranquility was returned to his church, the Webster Presbyterian Church in Houston, where every year since the historic landing in 1969, the church holds a special ceremony marking the anniversary of the moon's first communion.[12]

Four years prior to Apollo 11 in 1965, NASA was turning its attention from the one-man Mercury project to the two-man Gemini program, the lynchpin that would tie up all of the technological and procedural loose ends and make Apollo 11 possible. The program literally got off the ground on March 23, 1965, with the launch of original Mercury 7 astronaut Gus Grissom and rookie John Young aboard Gemini 3. By now, the rules as to what astronauts could carry in their PPK aboard a spacecraft had been formalized. Apparently, John Young didn't get the memo and an item he smuggled aboard the *Molly Brown*, Gemini 3's nickname, caused national heartburn.

During Mercury flights, astronauts were relegated to eating pureed food from tubes. When Project Gemini began, the menu was upgraded to cubes of freeze-dried food, which were coated in a gel to prevent them from crumbling and later reconstituted with water before being eaten. The morning of the launch, Wally Schirra went to Wolfie's Deli on North Atlantic Avenue in Cocoa Beach, Florida, and ordered a corned beef sandwich on rye, which he gave to Young who then put the sandwich into a pocket of his spacesuit.

According to Young, this wasn't the first time that a sandwich had been smuggled aboard a spacecraft. Without naming names, he wrote in his autobiography that it was in fact the third time.[13] During the mission, Young pulled out the sandwich.

"What is it?" Grissom asked.

"Corned beef sandwich," Young replied.

"Where did that come from?" the surprised Grissom asked.

"I brought it with me," Young replied nonchalantly. "Let's see how it tastes. Smells, doesn't it?"

Taking a tentative bite, Grissom noticed that once the bread was exposed to the spacecraft's high-oxygen atmosphere, it began to crumble into a grainy texture like corn meal, which littered the capsule with crumbs.[14]

"It's breaking up," Grissom commented. "I'm going to stick it in my pocket."

"It was a thought, anyway," Young said, defeated. "Not a very good one."

"Pretty good, though, if it would just hold together," Grissom said, reassuring the rookie.[15]

An innocent enough "experiment," right? Not to the press, who picked up on the story and ran with it. The story caught the attention of the House Committee on Appropriations in Washington, which quickly convened a hearing and demanded that NASA's Associate Administrator for Manned Spaceflight Dr. George Mueller, director of the Manned Spaceflight Center in Houston, Dr. Robert Gilruth, and administrator James Webb appear before the committee.

The hearing was a free-for-all. One congressman called the incident the "$30 million sandwich," and Illinois Congressman George Shipley said the incident was "just a little bit disgusting."[16]

"The sandwich was just a two-minute interval in the whole five hours and fifty-four minutes [of the flight]," Young wrote in his book. "It didn't even have mustard on it. And no pickle."[17]

The irony of this story is that only four years later, John Young was assigned to fly aboard Apollo 10. In addition to the normal freeze-dried fare, for the first time, the astronaut's menu also included individually wrapped slices of commercial bread and the fixings to make tuna, ham, and chicken salad sandwiches,[18] though no corned beef.

The tales of smuggled food and other contraband aboard spacecraft are generally one-off occurrences, with NASA strictly enforcing the rules and, for the most part, the astronauts playing by the book. By the time the United States moved from Gemini to the lunar missions of Apollo, the size of the PPK was increased from a six-by-seven-inch gray nylon drawstring bag to an eight-by-four-inch bag, giving the astronauts the freedom to bring more mementos with them.

On the Apollo 8 flight, Frank Borman carried a sphere of aluminum, which, when returned to Earth, was struck into 200,000 medallions that were presented to those who contributed to the Apollo program.

Alan Shepard on Apollo 14 famously brought two golf balls to the Fra Mauro formation on the moon. Using a makeshift golf club created by engineer Jack Kinzler (the same man who developed the flagpole that would make the American flag appear to blow in the breeze on the airless lunar surface), Shepard drove the balls into the largest sand trap ever. The club was a Wilson-Staff Dyna-Power 6-Iron golf club head attached to a lunar rock sampling scoop. Using a one-armed swing because of the bulky space suit, Shepard shanked the first ball into a nearby crater then drove the second ball across the desolate golf course, commenting that it went for "miles and miles and miles." There always has been speculation about how far that shot really went, but according to the US Golf Association (USGA), which used photos of the landing site taken by the crew before liftoff from the moon and high-resolution photos taken by more recent reconnaissance satellites, it is estimated that the ball traveled forty yards.[19] Not bad for a one-armed swing.

One controversial Apollo mission in the history of the PPK was that of Apollo 15. The three-man crew of David Scott, James Irwin, and Al Worden was the first to use the lunar rover and drive more than seventeen miles across the moon's surface. They also set the record (at the time) for the longest stay on the moon at eighteen hours and thirty-seven minutes. On August 2, 1971, near the end of Scott and Irwin's third extravehicular activity (EVA) on the moon, Commander Scott took out an item he had packed—a single falcon feather (Apollo 15's lunar module was named *Falcon*). Holding the feather in one hand, a geology hammer in the other, Scott turned to the television camera and dropped them both at the same time. With no atmosphere and virtually no gravity, the hammer and feather fell to the lunar dust at the same speed and landed at the same time, proving astronomer Galileo Galilei's theory of gravitational fields.[20]

Besides the falcon feather, the crew brought a few other items along for the ride, including some US currency, a few flags and medallions, and a friend's wedding ring. The mission commander had something else tucked away during the trip that NASA was completely unaware of and eventually would cost the crew of Apollo 15 their careers—four hundred stamp covers.

To a philatelist, covers are highly prized. Covers are cacheted envelopes with a unique image, stamp, or both that relates to an event in history, anniversary, or person. The postmark is what makes the cover special. For example, the Navy Department Library has a collection of such covers with photos of naval vessels on them and postmarks commemorating an important date in the ship's history, such as when the ship's keel was laid or when it was first commissioned.

In the event of a tragic accident, these covers proved particularly valuable to Apollo astronauts and their families. The astronauts found it impossible to obtain life insurance, so to provide for their families in case the unthinkable occurred, they took special envelopes with their mission logo printed on them, signed them, then had them postmarked at the Kennedy Space Center's post office the morning of their launch. The envelopes then were stashed away in safe deposit boxes, and if an accident occurred and the crew perished, the envelopes became highly valuable and fetched incredible prices at auction, which would sustain the families the astronauts left behind.[21]

NASA was not averse to sending such envelopes into space and did so on many flights. In fact, Apollo 15 carried with it 243 special envelopes in a NASA-approved OFK. However, as laid out in section 1214 of the eCFR, no one could benefit financially from the sale of these covers or of any other items flown into space. That's where the trouble began for the crew of Apollo 15.

Prior to the launch on July 26, 1971, a German stamp dealer, Herman E. Siegler, approached the crew with a business proposition. He would provide each of them with one hundred special envelopes that they would have postmarked the day of their launch at Cape Kennedy's post office, fly them to the moon, then return them to Earth, where they would be postmarked aboard the recovery ship and returned to Siegler to sell. The crew members would receive $7,000 each for their efforts, which they intended to save for their children. The crew agreed to the deal under the condition that they also would receive one hundred of the envelopes each, which they could sell on their own, and that none of the covers would be sold until the Apollo program was completed, which would be at splashdown of the Apollo-Soyuz mission in 1975. Siegler agreed and the crew made plans to use the money to set up a trust fund for their children.[22]

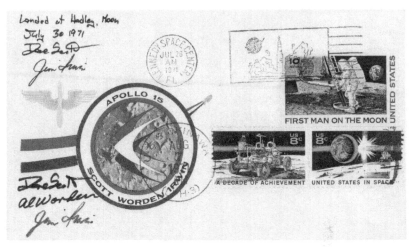

A "Siegler" cover canceled by Apollo 15 astronauts David Scott, Al Worden, and James Irwin.

At midnight on the morning of the launch, four hundred envelopes were delivered to the Kennedy Space Center's post office, where they were postmarked. The envelopes then were delivered to Scott before he boarded the capsule, where he stowed them away in one of the leg pockets of his spacesuit.

On the third EVA (the same one with the Galileo experiment), Scott took out one of the NASA-approved envelopes. On it was a newly minted stamp commemorating the tenth anniversary of US manned spaceflight. The stamp had two panels—a continuous image of the Earth rising with the lunar rover scooting across the moon. The stamps were inscribed with the words, "United States in Space . . . A Decade of Achievement."

On the moon, in the first ceremony of its kind, Scott pulled out a special stamp and ink pad and became a lunar postman as he postmarked the envelopes again. This mark imprinted the date and the words, "United States on the Moon." Scott actually stamped two of the envelopes because he had left moon dust fingerprints on the first one. These two envelopes would return to Earth and upon landing were postmarked a second time with the date and the words, "United States on the Earth" before being presented to the US Postal Museum.[23]

Meanwhile, the four hundred Siegler envelopes were still in Scott's pocket when he and James Irwin climbed back into the lunar module and left the moon for a rendezvous with Al Worden in the orbiting command module. Once docked, the moon walkers transferred themselves and their precious cargo of moon rocks and PPK into the command module, the hatches were closed, and the *Falcon* was jettisoned, sending it crashing into the moon. Unfortunately, the crew had forgotten to transfer one PPK bag—the one that held their friend's wedding ring, which was lost forever.

Stepping out of the helicopter aboard the recovery ship, USS *Okinawa*, the crew hightailed it to the ship's post office, where they had their secret four hundred envelopes postmarked. With both the date of launch and recovery stamped on them, they would become highly collectible. After returning to dry land, the three men divided three hundred envelopes among themselves and sent the remaining hundred to Siegler. When Siegler received the prized envelopes, the underhanded stamp dealer broke the agreement and immediately put the envelopes up for sale, netting a cool $150,000.[24]

Scott, Worden, and Irwin began to panic, realizing that this scheme could cost them dearly. Immediately upon hearing about what Siegler had done, they returned the money the dealer had given them and turned the envelopes over to NASA. Despite Deputy Administrator George Low's defense of the crew to the press, which emphasized the extreme stress that astronauts heading to the moon are under and that "their poor judgment in carrying the unauthorized covers must be considered in this light," the astronauts were reprimanded. Irwin announced his retirement, so no further action was taken against him. Scott was not assigned to any other flights, and Worden reportedly was "phased out" of the manned spaceflight program.[25]

"The government never said that we did anything illegal," Worden said in a 2011 interview with *Smithsonian Magazine*. "They just thought it wasn't in good taste."[26]

In 1983, Worden sued NASA for the return of the covers. Surprisingly, it was a short fight and the agency turned them over without a fuss. One rumor that has been tossed around to explain this sudden about-face was that earlier that year aboard the shuttle *Challenger* (STS-8), the crew carried thousands of similar but agency-authorized covers that were destined to be sold.[27]

But, as stated earlier, these tales of smuggling as a prank, an "experiment," as with the corned beef sandwich, or a chance to make a quick buck are quite out of the norm. For the most part, astronauts play by the rules, and over the years have taken some unique personal items into space in their PPKs. The crew of Apollo 11, for example, packed the usual flags and medallions and, as mentioned earlier, Buzz Aldrin took a chalice and vial of wine for communion. Command module pilot Mike Collins packed a shaving kit that included a razor, a tube of Old Spice shaving cream, and one unusual item—a hollow bean.

"Some things belonged to me," Collins told reporters, "and some to others. The only criterion was that the object had to be small, but none matched the ingenuity of one gent, for whom I carried a small hollow bean, less than a quarter-inch long. Inside it was fifty elephants, carved from slivers of ivory, which he planned to distribute to his coworkers after the flight."[28]

Mission Commander Neil Armstrong would be remembered forevermore as the first man to walk on the moon, a moment that would dramatically change the world as we know it, much like the moment when his fellow Ohioans, Orville and Wilbur Wright, gave man wings with their first airplane flight. A "devotee" of the pioneering aviators from Ohio, Armstrong was well aware of the significance of his small step and of how far mankind had come since the days of the Wright brothers. It had been only sixty-six years since the brothers first took flight at Kitty Hawk in the Wright Flyer I. As a way of celebrating this extraordinary advancement in human technology and linking that first flight with the giant leap he was about to take on the moon, the US Air Force Museum presented Armstrong with a piece of muslin fabric and small pieces of wood from the Wright brothers' first plane to take with him.

It's interesting to note that more than fifty years after Apollo 11, the Wright brothers again would fly into space, this time aboard the first helicopter to fly on another planet—*Ingenuity*. The helicopter landed on Mars in 2021 as a passenger aboard NASA's *Perseverance* rover. Onboard the tiny autonomous helicopter was another small swatch of fabric from the Wright brothers' plane. On April 19, 2021, *Ingenuity* took wing for its first flight in the thin Martian atmosphere. The grandniece and grandnephew of the Wright brothers, Amanda Wright Lane and Stephen Wright, said of the flight, "Wilbur and Orville Wright would be pleased to know that a little piece of their 1903 Wright Flyer I, the machine that launched the Space Age by flying barely one quarter of a mile, is going to soar into history again on Mars."[29]

Other items taken to the moon during the Apollo missions included a small silver statue of a man created by artist Paul Van Hoeydonck, which represented all of the astronauts and cosmonauts who had given their lives for the exploration of space. Apollo 15's Dave Scott placed the statue in the lunar dust along with a plaque that listed the names of the fallen.

In 1972, Apollo 16 astronaut Charlie Duke left his family on the moon—or at least a photo of himself, his wife, and two children. According to Duke, the idea was "just to get the kids excited about what dad was going to do. "I said, 'Would y'all like to go to the moon with me? We can take a picture of the family and so the whole family can go to the moon.'"[30] Albeit a touching gesture, time has more than likely taken its toll on the image, rendering the photo blank due to years of exposure to harsh cosmic elements. "[The photo] was very meaningful for the family," he said. "In the end, that's all that matters, right?"[31]

The PPK has played an important role in the American space program through the years as a reassuring piece of home in the hostile and hyper-stressful world of an astronaut. That tradition continues to this day aboard the space shuttle and now on the International Space Station as more people participate in long-duration spaceflights. Items launched into space continue to run the gamut, from the historic like a cargo tag from the first English settlement in America at Jamestown in 1611, to sports related like dirt from Yankee Stadium's pitcher's mound taken aboard the space shuttle *Endeavor* by astronaut Garrett Reisman, to pop culture memorabilia including Luke Skywalker's lightsaber from *Star Wars*. As mankind begins a new era of exploration with a return to the moon and, eventually, manned flights to Mars, it will be interesting to follow the astronauts and what they pack along "just in case" for their incredible journey into history.

11

WRECKED BY THE MOST
EXPENSIVE HYPHEN
IN HISTORY

In the final moments of the Mars rover *Perseverance*'s seven-month-long journey to the Red Planet during what is known as the "seven minutes of terror," Twitter accounts from the rover's team at the Jet Propulsion Laboratory (JPL) in Pasadena, California, flashed to life.

"Director of Planetary Science Lori Glaze and I have our #NASAJPL lucky peanuts in hand," tweeted Thomas Zurbuchen, NASA's associate administrator for the science mission doctorate.

JPL Chief Engineer Rob Manning posted a selfie of himself holding a jar of peanuts with the Mars 2020 logo emblazoned on it.

Then the mission's landing team joined in, tweeting, "We are ready! @NASAPersevere's entry, descent, and landing (EDL) team, excitedly tossing lucky peanuts. They've ensured the rover is best equipped for a successful landing on #Mars."[1]

With that, *Perseverance* hit the Martian atmosphere at 12,100 miles per hour, hitting a narrow entry window—a corridor, if you will—that, if missed after traveling and surviving its 126-million-mile journey, would end the mission, as the spacecraft would either be incinerated or skip off the atmosphere like a stone skimming across a lake, hurling the craft back into space. During the next four minutes, *Perseverance*'s heat shield bore the brunt of the friction caused by the craft racing through the thin Martian air, protecting the rover from the blazing 2,370-degree heat.

Seven miles above Jezero crater, the rover had slowed by 11,000 miles per hour, still faster than the speed of sound, when it jettisoned the heat shield and deployed a seventy-foot-wide supersonic parachute that slowed the craft to 160 miles per hour. The parachute and back cover were then jettisoned, and eight retro-rockets fired to continue to slow the descent and position the craft for touchdown.

Seventy feet above the ground, a crane aboard the spacecraft began lowering the rover. Only a few feet off the ground and traveling at two miles per hour, the rover *Perseverance* is cut loose from the crane and softly touches down in the Martian crater to begin its mission. Scientists, engineers, and peanuts made the heart-stopping but successful landing possible.

"Things are so complicated," Mars 2020 Deputy Project Manager Matt Wallace told the gathered press after the landing. "We're running a couple billion lines of flight software code. I think we had something on the order of 30,000 parameters to set and get them all right. You know, get through 10 or 12 Gs of deceleration, a supersonic parachute deployment, eight big main engines had to fire, our Terrain Relative Navigation Hazard Avoidance System had to perform as it was designed, and, you know, it's just never easy."[2]

"It's just never easy." Now that is an understatement if ever there was one. The fact is, of all the missions to Mars, nearly half either completely or partially failed during launch, on the journey, or during landing. So how—with all of the incredible scientific technology and know-how that makes such a mission possible—did peanuts play a role in the heart-stopping landing of *Perseverance*? It's a little-known tale that begins in 1958 with the first attempt to put a satellite in orbit around the moon and the series of failures that followed and demonstrated just how dangerous flying into deep space can be.

With the Soviet Union literally launching the space race with Earth's first satellite, Sputnik 1, in October 1957 and following the United States' explosive false start with the failure of Vanguard and eventual orbiting of Explorer 1 in 1958, it seemed impossible to think that either of the two fledgling space adversaries was in a position to send spacecraft to another world. The space age was merely an infant taking small, tentative, and painful first steps. But they did, and the United States was the first to shoot for the moon.

On March 25, 1958, the US Department of Defense announced that the Air Force's Ballistic Missile Division would launch a series of probes to the moon. The first of the series, Able 1, was a fiberglass cone-shaped satellite built by Space Technologies Laboratory that would send a magnetometer, a micrometeoroid detector, two temperature sensors, and a television camera

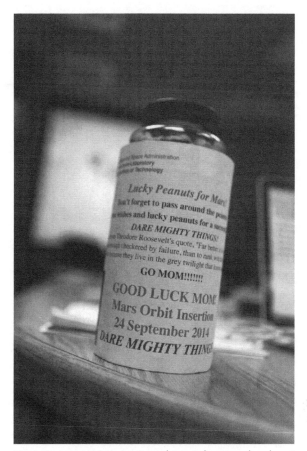

Navigators at NASA's Jet Propulsion Laboratory break out
"lucky peanuts" in support of India's Mars Orbiter Mission
(MOM) during its Mars Orbit Insertion. *NASA/JPL-Caltech*

into lunar orbit. This would be the first attempt by any nation to send a
spacecraft beyond Earth's orbit. The probe would be launched aboard the
Air Force's Thor-Able rocket, which was modified to have three stages and
used components from the Navy's Vanguard rocket.

On August 15, 1958, launch pad 17A at Cape Canaveral Air Force Station in Florida lit up as Thor's main engine ignited. The white rocket with
its long, slender nose began to slowly rise from the pad. Just 73.6 seconds
later, the mission was over. The rocket's main engine exploded, hurling the
tiny satellite into the Atlantic Ocean.[3]

All flights in the Able project were to be sequentially numbered—Able 1, Able 2, and so on. In subsequent years, the missions were renamed Pioneer, with Able 1 posthumously renamed Pioneer 0, making it the first in a series of lunar probes that NASA, the Air Force, US Army, and JPL would launch, and the first of a series of lunar failures. Two months after Pioneer 0, Pioneer 1 suffered a malfunction and never made its goal of lunar orbit. A month after that, Pioneer 2 reached an altitude of only 951 miles after launch before crashing back down and burning up in Earth's atmosphere. In fact, all nine Pioneer probes were unsuccessful with their missions. Pioneer 4 came the closest to completing its goal of performing a lunar flyby to take images of the surface, but instead of passing the moon at 20,000 miles, it passed at 37,000 miles, not close enough to take photos.

Malfunctions continued into 1964 with Project Ranger. The Ranger probes were designed to impact the moon, but on the way to their demise, several scientific instruments would relay data on cosmic rays and micrometeoroids near the moon back to Earth and televise images of the surface before crash landing.

When the Ranger program began, it looked like attempts to fly to other worlds were completely out of the hands of engineers, almost as if an unseen force was telling them, "don't do that." Ranger 1 and Ranger 2 were both destroyed after their Agena rockets failed to send them on their way to the moon. Both satellites fell short of escaping Earth's gravity and instead burned up upon reentry into Earth's atmosphere. Ranger 3 and Ranger 5's navigation systems experienced errors, causing them to completely miss the moon. Ranger 4 was completely disabled just after launch, and Ranger 6 completed its journey and crashed on the moon, but its television system failed en route.

By now, no one could blame engineers and scientists at NASA and JPL for believing that the American unmanned space program was jinxed. Despite the odds, they were going to try one more time with Ranger 7.

On July 28, 1964, the Ranger probe was readied for launch at Cape Kennedy. The engineers manning the consoles in the firing room were tense. What gremlin would pop up on this flight? Mission trajectory engineer Dick Wallace had the answer to calm the team's nerves.

"I thought passing out peanuts might take some of the edge off the anxiety in the mission operations room," he later recalled.[4]

The launch was perfect, and Ranger 7 had a near-flawless flight to the moon. On July 31, the probe began sending back 4,316 images of the moon's surface, taking the last image only 2.3 seconds before impact.[5]

Was it a coincidence that the first successful US flight to the moon occurred when the peanuts arrived? Maybe. But there is evidence—although not scientific in nature—that they do play an important role in the success of a mission. According to NASA, on a few rare occasions the peanuts were not on hand, the gremlins reappeared. During the subsequent launch of an unnamed probe, the peanuts were not in the control room, and shortly after liftoff, the spacecraft was lost. On another mission, the launch was scrubbed for forty days for various reasons. It wasn't until the peanuts made an appearance in the control room that the launch went off without a hitch. And the launch of the Cassini-Huygens probe that would orbit Saturn and land the Huygens probe on the planet's moon Titan was delayed due to damaged thermal insulation at the launch pad.[6] After the finding, engineers brought peanuts, and sure enough, there was a successful launch on October 15, 1997, and the mission completed successfully.

Superstitions aside, the odds of an unmanned probe traveling millions of miles from Earth safely to a distant planet are astronomical, and it's not only the severe conditions of space travel that cause a mission's untimely death. Such was the case with Mariner 1.

Mariner 1 was the first in a series of spacecraft developed by the United States for deep-space planetary exploration. Designed to explore Mars, Venus, and Mercury, these probes achieved several firsts, including the first interplanetary fly-by, the first spacecraft to orbit another world, and the first to use the gravity of a planet for "gravity assist,"[7] a method of using another planet to sling a spacecraft deeper and faster into space to rendezvous with yet another world. The program, however, got off to an inauspicious start only seconds after launch.

The evening of July 22, 1962, saw the launch pad at Cape Canaveral abuzz with activity as Mariner 1, sitting atop its gleaming silver Atlas-Agena rocket, was prepped for what was to be a historic flight, the first-ever fly-by of a distant planet, Venus. The spacecraft would take photos and perform rudimentary science experiments as it flew past Earth's nearest neighbor in the solar system.

At precisely 9:29 p.m., the main engines ignited, and the sleek rocket lifted off, illuminating the Florida sky. All systems were functioning normally as the rocket arced over the Atlantic Ocean. Suddenly, the vehicle began an unexpected yaw movement, twisting around its center of gravity. The rocket was out of control, and the range safety officer had no choice but to push the button and destroy the $80 million ($713 million in today's 2021 dollars) spacecraft only five minutes after launch. According to a NASA spokesman, "The booster had performed satisfactorily until an unscheduled yaw-lift (northeast) maneuver was detected by the range safety officer. Faulty application of the guidance commands made steering impossible and were directing the spacecraft towards a crash, possibly in the North Atlantic shipping lanes or in an inhabited area."[8]

An investigation into why the launch vehicle went awry found that it was human error. Buried deep in the millions of lines of software code, a single, lone hyphen had been left out. As NASA spokesman Richard B. Morrison explained, "[The hyphen] gives a cue for the spacecraft to ignore the data the computer feeds it until radar contact is once again restored. When that hyphen is left out, false information is fed into the spacecraft control system. In this case, the computer fed the rocket hard left, nose down [data] and the vehicle obeyed and crashed."[9]

It was a huge embarrassment for NASA. Headlines in the *New York Times* shouted, "For Want of Hyphen Venus Rocket Is Lost,"[10] and famed science-fiction author Arthur C. Clarke proclaimed that Mariner 1 "was wrecked by the most expensive hyphen in history."[11]

The short-lived flight of Mariner 1 was a hard lesson for NASA, but it was learned, and the remaining nine Mariner missions were successful. As for other unmanned NASA missions through the years, only one other flight experienced a similar human glitch. It was a bit of miscommunications that ended with the same result.

The Mars Climate Orbiter, or MCO, was the second of a trio of spacecraft that were part of the Mars Surveyor Program, a series of probes designed to further photograph the planet but, more importantly, to study its atmosphere and the polar ice cap. MCO was scheduled to arrive at the same time as the Mars Polar Lander and act as a relay station for data from the lander until the lander eventually shut down two months later, at which time MCO would begin its own scientific work.

On September 23, 1999, after nine and a half months of interstellar travel, all seemed well with the orbiter as its engines fired up once more to put the craft into a Martian orbit. As the spacecraft whizzed around the far side of Mars, tracking stations lost contact with MCO as expected and waited for it to reappear on the other side. It never did.

Once again, an investigation set out to find what caused the disappearance of the orbiter. It was determined that the navigation team at the Jet Propulsion Laboratory was using the metric system to calculate position. The company that built the spacecraft, Lockheed Martin, used the imperial system (pounds, feet, etc.). Someone had forgotten that the two units of measure had to be converted in order for one to understand the other. What that meant was while JPL was looking at pound-seconds for navigation, Lockheed was sending newton-seconds. The confusion caused the spacecraft to dip too low into the Martian atmosphere where it burned up.

Mars seems to be the worst offender when it comes to turning an ambitious scientific mission into a complete failure. Aside from the two manmade, mission-ending errors just mentioned, mechanical issues, interplanetary travel, and the planet itself seem to gobble up spacecraft at an incredible rate. Of the forty-eight international missions to the Red Planet listed on NASA's Mars Exploration website, thirty-two have failed.[12] Some of those missions are considered partial successes, like Britain's Beagle 2.

The barbecue grill–sized Mars lander, Beagle 2, was scheduled for a Christmas Day 2003 landing in a flat, sedimentary basin called Isidis Planitia to look for signs of life on Mars. Riding aboard the European Space Agency's Mars Express Orbiter, the little lander successfully deployed, but almost immediately, all contact with the tiny craft was lost and it was presumed to have crash-landed. Ten years later, high above the planet, NASA's Mars Reconnaissance Orbiter spotted something unusual on the surface. The orbiter spotted it once again in 2014, and after careful analysis, it was determined that the object was indeed Beagle 2.[13] The little lander that could had successfully landed on Mars. It just couldn't phone home.

One of the most spectacular failures was the European Space Agency's Schiaparelli spacecraft. The lander was named for Italian astronomer Giovanni Virginio Schiaparelli, who began making maps of Mars in 1877 detailing *canali*, or canals, on the planet, which set off a firestorm of speculation that there was intelligent life on the planet who were responsible for building the canals.[14]

The plan for the Schiaparelli lander was for it to parachute to the ground, but three minutes after it first encountered the Martian atmosphere, things began to go horribly wrong. On October 19, 2016, the spacecraft deployed its parachute and jettisoned a protective back shell covering. Without warning, the lander began spinning at an incredible speed. The guidance, navigation, and control systems were overwhelmed with data, causing its computers to misjudge its altitude. Believing the spacecraft was below ground level (you read that right—the computers thought the craft was underground), the computers released the parachute prematurely. The landing thrusters were fired for only three seconds instead of the programmed thirty seconds, but even so, it was a futile effort because Schiaparelli wasn't even close to being on the ground. It was actually 2.3 miles in altitude and careening toward the surface at an estimated 335 miles per hour. The impact created a new crater on the Martian surface that was later photographed by NASA's Mars Reconnaissance Orbiter. The new crater was aptly named Schiaparelli crater.

The Russians are not immune to Martian gremlins. In fact, the Soviet Union and Russia are responsible for most of the failed missions to Mars. One of those was an ambitious joint mission between Russia and China that would deploy the first Chinese interplanetary satellite, Yunghuo-1, then send a Russian probe to the Martian moon Phobos, where it would land, collect samples of the soil, then return them to Earth. The spacecraft never made it out of Earth's orbit due to its engines failing to ignite.

However, it wasn't Mars that dealt the biggest blow to the Soviet Union's space program, but an unmanned test of the booster rocket designed to beat the Americans and land the first man on the moon. It was a test that ultimately ended the space race to the moon.

It was the summer of 1969. The United States and Soviet Union raced to put the first human on the moon. With every new mission following the first tentative steps into space by Yuri Gagarin and Alan Shepard, each country upped the ante with more daring space firsts—the first man in space, the first spacecraft rendezvous, the first spacewalk, the first docking of two spacecraft, even sending the first men a quarter of a million miles from Earth to the moon, where they came within sixty miles of the lunar surface as a dress rehearsal for the big event, the first lunar landing.

As the dreams and hard work of Wernher von Braun, NASA employees, and the many contractors who made the moment possible were at the threshold of reality, halfway around the world, the unimaginable was happening—the Soviet space program was literally crashing and burning.

It was an inconceivable event considering the string of space firsts the Russians had racked up early in the space race. They first showed their scientific prowess in space by launching and orbiting the first artificial satellite, Sputnik, in 1957. The tiny basketball-sized orb had caused worldwide panic. After all, if the Russians could orbit a satellite, then they could surely launch a nuclear bomb into space and detonate it anywhere in the world.

From that moment on, the Soviets tallied up an impressive list of space firsts including orbiting the first living creatures—a pair of dogs and the first human in 1961. Two years after Sputnik, the Russians took an even bigger leap in the race for the moon with their Luna missions.

Luna 1 was launched in early 1959, becoming the first man-made object to do a fly-by of the moon, passing by at 3,725 miles above the surface. On September 14 of that same year, Luna 2 became the first man-made probe to reach the surface of the moon. Although it was a crash landing at the Sea of Serenity, it was nonetheless an impressive feat. Wrapping up the year, Luna 3 took the first photographs of the far side of the moon.

Phillip Morrison, professor of physics at MIT, once said of science that "the probability of success is difficult to estimate."[15] When it came to the race for the moon, truer words were never spoken. When the Soviets burst out of the starting gate of the space race, it seemed a safe bet that landing the first man on the moon highly favored the Soviet Union. Still, the United States persisted and scored its own spectacular space firsts in the early 1960s, quickly narrowing the lead. By 1969, the odds had swung in favor of the Americans, especially after they had successfully orbited not one but two crews of astronauts around the Earth's satellite in preparation for the Apollo 11 mission. With the odds changing so rapidly, what happened next to the Soviet space program during the first week of July 1969 looked like a gallant last-ditch effort to beat the Americans, but in reality, the opposite was true.

As different as the two space programs were, they followed a remarkably similar track to prepare for a moon landing, both working slowly and methodically to build the necessary technologies and databases of experience and knowledge that would make the dream a reality. Although the science, training, and scheduling of both programs looked similar, when it came to the hardware and actual nuts and bolts of a flight plan, all similarities ended.

Take, for instance, the LK (lunar craft), the vehicle the Soviet Union designed to land men on the moon. The concept and design were similar to the American lunar module (LM) with three big differences. First, Soviet scientists were concerned about the weight of the vehicle, so instead of having two cosmonauts land on the moon, they constructed the vehicle to carry only a single person, whereas the US LM carried two.

Another difference was that the LM was a two-stage machine with both an ascent and descent stage. The descent stage powered the lander to the lunar surface. It also held extravehicular activity (EVA) gear, which on later Apollo missions included a lunar "golf cart" and eventually the lunar rover. The descent stage also acted as a launch pad for the ascent stage, which returned the astronauts and their precious cargo to the orbiting command module (CM) for the return ride home. The Soviet LK, on the other hand, was a single unit with only one engine that both landed the craft and powered it back into lunar orbit.

Finally, the most significant difference between the two craft was how they returned to their orbiting ride home after leaving the moon. The LM physically docked with the command module, which allowed the astronauts to travel in relative safety between the two craft through a docking tunnel. The LK, on the other hand, did not dock with its return spacecraft, but only rendezvoused with it, floating in close proximity. The lunar cosmonaut, who, one would assume, was exhausted after the stress of landing on the moon, exploring by himself, and launching back into orbit, had to climb out of the lander and take a perilous spacewalk a quarter of a million miles from Earth to climb into the orbiter.

The Soviets had a lander but now needed a delivery vehicle to take it out of Earth's orbit and on to the moon. Making such a trip required a massive rocket. For the United States, Dr. Wernher von Braun and his team designed the mighty Saturn V, while halfway around the world, the Soviets designed and built the N-1.

Measuring 344 feet, this behemoth of a rocket was only nineteen feet shorter than the Saturn V. The rocket's first stage was powered by thirty engines. The rocket weighed more than six million pounds at launch and could carry more than 200,000 pounds of man and machine into low-Earth orbit and 58,000 pounds beyond Earth's gravity to the moon.

In February 1969, the stage was set for the first test of this massive machine. At 12:18 p.m. Moscow time, the towering rocket's engines were ignited. Ever so slowly, it lifted off the pad, cleared the launch tower, then began picking up speed as it roared into the sky. Two minutes later, the rocket's first-stage engines began to falter, and the rocket plummeted back to Earth with a mighty explosion.

The second launch attempt occurred on July 3. Only weeks before Apollo 11 would be launched, hopes for the Soviet Union to send a man to the moon rested on a successful flight of this N-1. The rocket was equipped with a Zond spacecraft that would one day orbit cosmonauts around the moon much like the American command module. The version of the Zond for this test flight was stripped down and equipped with cameras so that it could photograph possible lunar landing sites for manned missions. Also onboard the N-1 was a dummy LK lander.

At precisely 23:18:32 Moscow time, the engines of the first stage ignited and the rocket began to free itself from the bonds of Earth's gravity, easing away from the launch pad. Ten seconds later, the rocket burst into flames, destroying not only the rocket, but also the launch complex and surrounding ground facilities as well. Remarkably, the Zond spacecraft was tossed from the explosion and was found intact a half mile away.

An investigation into the accident revealed that a single bolt had come loose and was sucked into an oxygen pump, causing an automatic shutdown of its engines. In an instant, the N-1, along with the Soviet's goal to reach the moon before the Americans, was destroyed. With NASA preparing for the Apollo 11 landing, the Soviets made the decision to scrap their own manned lunar landing objectives for the time being and instead began focusing on a low Earth orbit program of scientific studies, ultimately designing the first space station to orbit the planet.

12

SPACE IS OPEN FOR BUSINESS

A grainy image of a space station flickered on television screens across America as an anonymous voice announced, "We have a spectacular view of the Mir space station over the South Pacific." The scene quickly cut to the inside of the station, where Commander Anatoly Solovyov and flight engineer Pavel Vinogradov are seen floating weightless.

Amid the Russian chatter with Mir's control center in Kaliningrad, an American voice drops in to greet the pair, who are circling the Earth more than two hundred miles above. "Hello! This is Dave King in New York." There is no response. The two cosmonauts fight through the spaghetti-like tentacles of headphones that float serenely around them in zero gravity, trying to find a headset that will allow them to communicate with King.

Viewers watch as the two put on a second set of headphones. "The beauty of live TV," King interjects, filling dead air until he can talk with the space travelers. "See everything floating? Isn't it cool?" A live audience watching the spectacle chuckles. Cut to the studio, where we see the host of the show, Dave King, watching the cosmonauts on a small thirty-inch portable television monitor. "The pen that they are looking at right now," King begins, "is our first product in the show of *Extreme Shopping*. Now, it is called the Fisher Space Pen, and they will be showing you how it works."[1]

The date was February 7, 1998, and Solovyov and Vinogradov were live on the cable shopping channel QVC hawking the Fisher Space Pen, the world-famous writing instrument that could write upside down even in weightless conditions. The pen normally sold for $36 on QVC, but on this edition of the *First Friday Extreme Shopping Show*, it could be purchased for only $32.72 plus shipping and handling.

The communications glitch continued for a few more minutes as King kept the show rolling. On the monitor next to him, viewers could see one of the Mir crew members dip below the shot. When he reappeared, he held up a white card on which he had written the abbreviation "QVC" with one of the pens. It was the first live televised space-oriented shopping event in history and a moment that brought the marketing of space to the public's consciousness.

It was evident that space was now open for business, but this wasn't the first time that space and Mir in particular had been used as a marketing tool for products and certainly would not be the last. Through the years, everything from food products to digital picture frames to electric cars have been marketed in space, and the trend continues moving forward unabashedly at full steam.

There are two avenues that a marketing firm can take when connecting its product with space exploration. The first is direct marketing, purposefully using anything and everything related to space as a vehicle to promote the product. In this instance, you need a willing partner to help create the advertising. NASA has a rule on the books that prohibits commercial companies and NASA employees from earning money off of the US space program through advertising or promotional work. For the Russians, advertising has been a source of income that for many years helped keep their space program going, and that was the case with the cosmonauts' appearance on QVC. But more on Soviet and Russian marketing in a moment.

The other type of promotion is unintentional, wherein circumstances and timing bring a product and manned spaceflight together, giving the space program a needed piece of equipment, with the added advantage of boosting sales and public awareness of the product without it actually being sanctioned by NASA. That is what makes the story of the Fisher Space Pen so interesting. Its inventor, Paul C. Fisher, had no intention of marketing the pen as a space product. He was simply looking to create a better pen but being seen as a space-age spinoff didn't hurt.

Fisher worked in an airplane propeller factory during World War II then, following the war, moved on to work at a small pen factory. It was there that he realized that pens of the day were not user friendly by any means. Ink ran out of the tip of the pen. Pens quickly dried out, rendering them useless. The ink easily smeared. Once on paper, it took the ink forever to dry. In fact, a person could press his thumb onto a signature and transfer an image of that signature to another document up to a week later.[2] With that in mind, Fisher set out to create a better pen.

During the ensuing years, Fisher experimented with thousands of combinations of ink and cartridges from which to dispense it. He tried simply sealing the cartridge, making it airtight. He tried pressurizing the cartridge, which worked to a point, but the ink was still unmanageable. According to the inventor's son, Cary Fisher, his Dad dreamed of his grandfather, who told him that by adding a resin to the ink, his problem would be solved.[3] Paul Fisher searched for just the right resin when he came upon one used by the rubber company, B. F. Goodrich. Fisher added a small amount of it to ink, which made it more viscous and stickier.[4]

Now that the ink issue was resolved, it was back to the cartridge. Fisher concocted a hermetically sealed chamber to hold the ink that was pressurized with nitrogen. After years of experimenting and investing $1 million into the project, he had invented what he believed was the perfect pen. It could write in temperatures ranging from minus 30 degrees Fahrenheit to more than 250 degrees. It could write upside down. It could even write under water. But it wasn't called a "space pen." In fact, Fisher had no intention of using the pen in space. It was simply a better pen to use here on Earth. He called it the Anti-Gravity Pen, or AG7, and patented his invention in 1967.[5]

Until this time, the American manned space program had been using mechanical pencils to write with, which caused incredible heartburn for NASA. The agency had procured thirty-four units of mechanical pencils from Tycam Engineering Manufacturing at a cost of $4,382.50. That means that a single pencil cost $128.89, an exorbitant price to say the least. The public and Congress thought it was a frivolous purchase, forcing NASA to back out of the deal.[6]

Seeing that NASA couldn't find a pencil to write with, the Fisher company offered the agency its pens to use on the upcoming Apollo missions to the moon. Needless to say, after the pencil brouhaha, the agency was

Though not flown aboard the Apollo or Soyuz spacecraft during the joint US-Soviet Apollo-Soyuz Test Project mission in 1975, the flight was still marketed on the ground with the release of commemorative Apollo-Soyuz cigarettes, a pack of which is on display at the US Space and Rocket Center in Huntsville, Alabama. *Joe Cuhaj*

reluctant to make another misstep but intrigued by the concept of the pressurized pen. NASA began eighteen months of rigorous testing, and in 1967, the agency signed off on the deal and purchased four hundred pens at $6 per pen. With the blessing of NASA, the AG7 was renamed the AG7 Fisher Space Pen,[7] and though the pen was already popular around the world, promoting it as the pen used by American astronauts enhanced sales incredibly.

It wasn't only the US space agency that purchased the pens. In February 1969, the Soviet Union purchased one hundred of the pens. Until that point, cosmonauts had been using grease pencils to write with. Even the Chinese space program now uses the Fisher Space Pen.

To mark the fiftieth anniversary of the invention of his fledgling company's very first pen, Fisher designed and marketed a special fiftieth anniversary edition, one of which is on permanent display at the New York Museum of Modern Art for its "elegant American design."[8]

Another company that didn't really need a boost from the American space program but nonetheless benefited greatly from it was the Omega

Watch Company. The small Swiss watchmaker opened its doors in 1848 as Louis Brandt and Fil in the town of La Chaux de Fonds. Its founder, Louis Brandt, quickly made a name for himself and his company around the world with handsome and extremely precise timepieces. It wasn't until 1885 that the company created its first mass-produced watch, the Omega Caliber, and changed the company name to Omega.[9] By the turn of the century, Omega was recognized worldwide as a premier watchmaker, even becoming the official timekeeper of the Olympics beginning with the 1932 games in Los Angeles, California.[10]

Following the success of its underwater diving watch, the Seamaster, Omega horologist Claude Baillod created the first Speedmaster in 1957. In addition to the standard clockface, the watch sported a large third hand that acted as a stopwatch and the now-famous triple dials—one as a secondary second hand, another as a minutes-elapsed counter, and the third as an hour-elapsed counter. The first time the watch was used by an American astronaut was during Wally Schirra's Mercury flight aboard *Sigma* 7 in 1962, when he wore his personal Speedmaster on the mission.

Following Schirra's flight, NASA began a search for the best watch with which to equip astronauts. Not wanting to spend money and time developing its own since there were already outstanding watchmakers in the world, the agency asked ten companies for sample watches without telling them what they wanted them for. Of the ten, only four sent samples—Rolex, Longines, Hamilton, and Omega.

The watches were sent to former NASA project engineer and program manager James H. Ragan for evaluation, whereupon they underwent a grueling battery of tests to see if they could operate in temperatures between 160 degrees and 0 degrees Fahrenheit, relative humidity of 95 percent, a pressurized oxygen atmosphere, acceleration to 7.25 Gs within 333 seconds, ninety minutes of decompression from a vacuum environment, and excessive vibration. Years later, Ragan recalled the testing process:

> Hamilton's was a pocket watch and a big sucker, so we threw that out straightaway. I bought three of each of the others: one for the astronauts, one for testing, and one for backup in case we did something wrong. The Rolex and Longines basically crapped out on the first test. We did eleven environmental tests and I just had to hope the Omega made it through the rest. It did, and the astronauts preferred the Omega, too. After every flight I'd take

their watches away for maintenance and they'd get pretty upset about it. They always flew with the same serial numbers. It was a lucky charm.[11]

The Omega Speedmaster watch became a critical piece of equipment for the men who landed on the moon. After landing, astronauts suited up then started the watch that clung to their wrist with a long Velcro strap. When Neil Armstrong became the first man to walk on the moon during Apollo 11, he left his Omega onboard the lunar module (LM) *Eagle* because the timing system in the LM was not functioning correctly. When Buzz Aldrin followed Armstrong onto the surface nineteen minutes later, he was wearing the first Omega on the moon. Sadly, a few months after the crew's safe return to Earth, Aldrin's watch was stolen and has yet to be recovered.[12]

The head of Omega's Brand Heritage Department, Petros Protopapas, recalls talking with the last man to walk on the moon, Apollo 17's Gene Cernan, who related an emotional story:

[Cernan] was asked what the Speedmaster meant to him. He became a bit emotional and said when he was on the moon closing the mission, he looked up and saw our planet and realized that, at the moment, he was the most solitary human being alive. Then he looked at his watch, and it made him imagine what was happening at home—what his wife and daughter were doing. This was the only thing he had to connect him to home.[13]

The world was enamored with the watch that went to the moon and the Speedmaster was renamed the Speedmaster Professional. Following the conclusion of the Apollo program and the safe return of Apollo 17 to Earth, Omega ran a full-page ad in newspapers that shouted in big, bold letters, "On the occasion of the successful completion of the Apollo program, Omega would like to say, 'Thank you, NASA!'"[14]

And what about that most famous (at least to kids who grew up in the 1960s) of NASA "spinoffs," Tang? You guessed it. It was another product that saw sales rocket (no pun intended) after its selection as a water alternative drink aboard the Gemini and Apollo missions. And, of course, with a little help from some creative Madison Avenue marketing.

Tang Breakfast Drink was invented by Dr. William Mitchell in 1957 for General Foods Corporation. A mix of chemicals, yellow food coloring, orange flavoring that was "packed with vitamins," his orange-flavored creation was originally called Tang Flavor Crystals. The crystals made their debut in 1959, hitting store shelves with a marketing campaign that the company hoped would speak to health-conscious moms. It was a healthy drink that "you don't squeeze, unfreeze, or refrigerate."[15] The campaign, and Tang, landed with a thud with meager sales at best. Enter NASA and John Glenn.

The Marine fighter pilot had a stellar career at the end of World War II and during the Korean War. The world had its first glimpse of the future astronaut in 1957, when he was paired with schoolboy Eddie Hodges on the television show *Name That Tune*.[16] It wasn't the TV appearance that made Glenn a household name; it was his ride aboard *Friendship 7* on February 20, 1962, and his nearly five-hour, three-orbit flight around Earth, the first orbital flight by an American.

During the flight, Glenn was tasked with performing some food experiments. You have to realize that up to that point, the two previous Mercury missions had been only fifteen minutes long, not enough time for even a snack. Scientists were not sure how the human body would take to eating while weightless: Could they swallow food and drink? Could they digest it? What was the best packaging and form of delivery? With Glenn's extended flight, it would be the perfect opportunity to perform these experiments.

Glenn was provided with several samples of food and drinks to try out. One of the drink mixes was Tang, which he drank from a pouch. The test was successful and afterward, Tang was packaged by NASA and sent into space with every manned spaceflight through the Apollo-Soyuz mission and into the early shuttle flights.

With Glenn's return to Earth, General Foods began blitzing television screens and newspapers with advertising about the space-age drink. Even though NASA did not develop Tang, the blanketed advertising made it seem as if it did, and sales of the drink mix made it one of the most popular of the 1960s.

What did the astronauts think of Tang? The outspoken second man to walk on the moon, Buzz Aldrin, wanted the world to know. In 2013, Aldrin was to present an award during *Spike TV's Guys Choice Awards*. During taping of the show, Aldrin tweeted: "Want to know what I think of Tang? Watch the #GuysChoice on @SpikeTV." During taping, he proclaimed, "Tang sucks." Later he told reporters that the comment was just a joke, and he loved the drink.[17]

NPR decided to do its own impromptu and unscientific poll of Americans to see what they thought of the drink; 57.1 percent of those who responded hated it, 29.43 percent thought it was great, and 13.47 percent had no idea what it was.[18] Despite those numbers, Tang is still with us today.

Getting back to QVC's marketing of the Fisher Space Pen live from Mir: the pen was not the only space-related merchandise hawked that cool February evening from the shopping channel's studios in New York City. Later in that same segment, cosmonaut Alexander Lazutkin, staged in Manhattan's famous Catch a Rising Star nightclub, would act as spokesmodel to demonstrate the next item for sale. Item number X-1098 was an actual Soviet Sokol KV2 spacesuit. The price tag—only $25,000. QVC did not accept Easy Pay.

Lazutkin demonstrated the suit for the cameras, pointing out the air valves and the canvas boots. On the shoulder, the flag of the former Soviet Union was embroidered. The cosmonaut verified that the suit was in excellent condition and made sure to mention that the suit could "even be used underwater."[19] And as a bonus, QVC was also selling small pieces of Mars rocks for anywhere between $90 and $2,500. These rocks were pieces of meteorites that were authenticated as being from the Red Planet.

By the end of the evening, 11 viewers purchased pieces of Mars, 530 bought a Fisher Space Pen, 6 callers made "serious inquiries" about the suit, and the Russian space program netted an undisclosed amount of money.

Why did Russia go all-out to do space marketing on the capitalist QVC channel? It was a matter of funding and a way to keep its floundering space program and the Mir space station aloft. The head of the Russian space program at the time, Yuri Koptev, made it clear to the world that Mir would be used on a regular basis as an advertising prop to keep the program flying.

"It doesn't make any difference for us what to advertise—cars or foodstuff," Koptev told reporters after the QVC segment. "The only condition is that advertising doesn't contradict legal and ethical norms."[20]

Mission Control Deputy Director Viktor Blagov added, "We advertise nonalcoholic drinks, sports gear, and many other things." Blagov went on to point out that American astronauts are public servants and prohibited from doing any type of promotion. "But we have no such law. [The lack of funds for the space program] forces us to take any steps, regardless of how strange they might seem."[21]

During the late 1960s, the R. J. Reynolds Company made good use of advertising space on matchbook covers, offering people the chance to own a Fisher Space Pen by sending in two empty packs of Salem cigarettes and $1. *Joe Cuhaj*

It was once shocking to see product placement on a spaceflight; after all, this was serious business. There was no time for mucking about with marketing and advertising. Today, long after Mir met its fiery fate, burning up in Earth's atmosphere after years of successful service, advertising in space appears to be the new normal, at least for the Russians, who seem to be running the market.

In 1997, Russia filmed the first television commercial in space.[22] The ad, which was filmed by future QVC star Alexander Lazutkin aboard Mir, begins with a frantic mission control in Moscow trying to regain contact with the space station. Applause erupts in the room when video communications are reestablished. The mission director, taking a long swig from an ice-cold glass of milk, asks the weightless cosmonaut Vasily Tsibliyev if there is anything he needs, to which Tsibliyev replies, "a glass of real milk. Like yours."

"Real milk in space?" the director replies. "It's never been done before."

The cosmonaut, staring out of a porthole at the blue Earth below suggests, "Why not Israeli milk?"

Then we learn how Tnuva's Long Life Milk is painstakingly tested in a Moscow laboratory then flown to Mir in a cargo ship, where the cosmonaut is seen happily sucking down a floating globule of the milk from a carton.[23]

Two years later, Pizza Hut entered the race for ad space (pun intended) by entering into a contract with the Russian space agency Roskosmos. The deal gave the pizza giant a thirty-five-foot-tall billboard—the side of a Proton rocket that was destined for the International Space Station—on which the company could paint its logo on the side. The fleeting few seconds of advertising netted the space agency somewhere between $1.9 and $2.5 million.[24] The pizza chain followed that up with another space first: the first pizza delivery to space. It was a six-inch salami pizza (pepperoni, apparently, failed miserably during tests), proving the slogan once and for all: "no one out-pizzas the Hut." It's a good thing the company didn't brag that its pizza is delivered in thirty minutes or the next one is free.

Others have since joined advertising's final frontier, including Radio Shack, which suggested in 2001 that viewers should give dad a digital picture frame for Father's Day by showing cosmonaut Yuri Usachev receiving a delivery of his special Radio Shack gift while in orbit. Even the world leader in ramen noodle production, Nissin Foods, got into the act when it introduced its microgravity friendly Space Ram in a 2005 Mir-based commercial.

Of course, any discussion of marketing and ad placement, whether earth-bound or otherworldly, would be incomplete without a word or two about the undisputed kings of advertising, Coca-Cola and Pepsi.

A trip into space is not only fraught with any number of dangers, but astronauts and cosmonauts have to make many sacrifices as well. Space can be a lonely place. Astronauts and cosmonauts are separated from the things they love, like family, friends, and the simple comforts of home. When it comes to the latter, the US and Russian space agencies do everything they can to provide their crews with a little piece of home to keep them connected and to make the journey and their mission more comfortable. NASA allows astronauts to take mementos from home with them in personal preference kits (PPKs), but one thing that can make a person isolated from the rest of the world feel more at home is good old-fashioned comfort food. The problem is, bringing such meals and treats into space has always been a challenge for a number of reasons. Leave it to the soda giants Coca-Cola and Pepsi to tackle the problem and to take the cola wars literally to new heights.

It all began simply enough. The Coca-Cola company wanted to experiment with developing a delivery system for carbonated beverages in space. How would a carbonated drink react in zero gravity? How could an astronaut drink from a can with sticky sweet liquid floating aimlessly out of it? What kind of mechanism could be developed to extract soda from a can but only when the astronaut wants a sip?

These were serious questions that NASA happily entertained. The comfort of their crews was always at the forefront of each mission, so the agency approved Coke's request to try an experiment with a new soda can on a shuttle flight. Coke's engineers went to work spending $250,000 to develop a version of its classic can with the iconic "wave" logo. Inside the can, a laminated plastic bag contained the drink and a carbon dioxide–pressurized bladder propelled the drink out.[25] The top of the can had a plastic mouthpiece that acted as a spout that astronauts could activate with their lips to get a swig of the "real thing."

The test of the new can was scheduled to fly on STS-51F, the eighth flight of the space shuttle *Challenger*. There was nothing unusual about the experiment, which is all it was. Just an experiment. NASA did not receive any advertising dollars from it and no promotional deals were struck. In fact, the experiment most likely would have gone largely unnoticed had it not been for the Pepsi Cola Company getting word of it and asking NASA to include its own dispenser on the same flight.

NASA officials agreed but this put the agency and astronauts in a bind. There was a long-standing policy that American spaceflights were not to be used for commercial purposes. With the ongoing earthbound cola wars continuing on the ground, the public could perceive these experiments as breaking that rule. The commander of the mission, Gordon Fullerton, and his crew reluctantly agreed to go forward with the experiment when it was given an official name: the Carbonated Beverage Container Evaluation experiment, or CBCE. The crew was relieved that it would remain an experiment and not become too commercial. However, Mission Specialist John David Bartoe had reservations. "I thought it was frivolous and detracting from the science of the mission," Bartoe told CollectSpace.com.[26] "I said, 'I'm not going to do it. I think it's a terrible idea.'"

Once NASA was onboard and all participants were satisfied with the arrangement, Pepsi sprang into action and quickly developed its own dispenser just in time for the July 29, 1985, launch. The can was less high tech than the Coke version and reportedly cost $14 million to develop.[27] It was a modified version of the type of can used to dispense spray cheese like Easy Cheese. The can was not pressurized but used chemicals to produce the carbonation.

Fullerton decided that to eliminate any appearance that the experiment was a taste test, the experiment would not be scheduled. Instead, the tests would take place only during the crew's downtime and away from cameras. Three astronauts would try the Coke can, three would test the Pepsi dispenser, and the seventh crew member, Bartoe, who had aired his grievances about the experiment, was "banned to the flight deck" while the experiment took place.[28]

The results were kept under wraps until after the flight, and even then, the crew did not comment on the taste of the beverages, only on the method of delivery. Mission Specialist Anthony England commented that the cans worked as expected. "When we opened them," he told the media gathered at a press conference, "they did not explode or anything of that sort. And the drink was immediately available upon request."[29]

The real problem was not the cans but the beverages themselves. In a weightless environment, the carbon dioxide that gives soda its fizz is not released when you open a can like it does on Earth, which means that the gas stays in the drink, causing the astronaut to ingest more gas than normal. After ingestion, the gas does not separate in the digestive system like it does on Earth, which means the astronaut experiences "wet burps," in which a little bit of liquid is expelled along with the gas, much like acid reflux.[30]

Despite NASA's best attempts at downplaying the experiment and without indicating the astronauts' soft drink preferences, the marketing wizards at both companies waited anxiously for any clue as to which can was tested first. It was eventually revealed that it was Coke, and the company's promotional department went into high gear creating advertisements touting that aspect of the tale. A print ad was quickly released showing a gloved hand with the starry background of outer space reaching for a can of Coke. The caption read, "For the first time in history, our astronauts carried soft drinks with them as they orbited the Earth. Coca-Cola, the first choice in refreshment around the world, is now the first drink tasted in space."

The Pepsi Cola Company downplayed the ad. "It's about as relevant as who's first in the birth of twins," spokeswoman Becky Madeira told *Forbes* magazine. "If they were first to be tested, it was the new Coke, and you can be sure it had to be washed down with a Pepsi."[31]

Now, don't get the impression that Pepsi was finished with marketing in space. PepsiCo's Russian division was reportedly toying with the idea of launching an "orbital billboard" consisting of a cluster of small satellites flying in formation. Orbiting at 250 miles above Earth, the satellites' mylar coating would reflect sunlight and feature a blinking logo promoting Pepsi's Russian energy drink, Adrenaline Rush. Shortly after the story broke on the Futurism.com website,[32] a PepsiCo spokesman emphatically denied that they were going ahead with the scheme.[33]

And that Coke commercial that aired at the beginning of the 2021 Tokyo Olympics that portrayed an American astronaut and Russian cosmonaut floating in the International Space Station, chasing globules of the soft drink after watching a US-Russian hockey game on TV? The internet was abuzz with rumors that it actually had happened. Spoiler alert: it was fictional. Sorry.

As the final frontier opens its doors to more commercial enterprises, you can be sure that marketing departments already have their heads in the stars planning the next, even more stunning, promotional stunt in space. It will be hard to top the Tesla automotive company, which amazed the world with the launch of Starman, a mannequin in a space suit seated behind the wheel of a convertible Tesla Roadster. Starman has already whizzed past Mars and will be circling the Sun every 557 days. A remarkable billboard if ever there was one. And the European Space Agency was looking ahead when it released a guide not long ago titled "Space Marketing: A New Programme in Technical Education."[34] Just a little sales training manual to get public relations departments ready for the new market.

Apparently when it comes to promotion and marketing, the stars are not the limit.

13

SEX IN SPACE IS MORE THAN JUST A BIG BANG

A group of astronauts, NASA scientists, and officials sat at a long table. A microphone was positioned before each of them as they fielded questions from the media, all the while fiddling with the mic stand and attempting to answer the questions to the best of their ability. Then it happened. The inevitable question that had been asked a million times before and that receives a titter of laughter from the gathered journalists each time. Without even seeing the panelists' faces, you can sense the eye rolls and expressions conveying "not again"; others nervously tap on the table hoping that they won't be the person to answer it.

Arguably, the most frequently asked question of astronauts and scientists is, "how do astronauts go to the bathroom in space?" A close second is the equally "pressing" question that the general public and the media press corps frequently ask without blushing: "Has anyone ever had sex in space?"

Spoiler alert: the answer is no. Whenever asked by reporters, NASA flat-out denies that any out-of-this-world hanky-panky has taken place. Even the Russian space agency Roscosmos emphatically denies any romantic interludes in space. As the deputy director of the Russian Institute of Biomedical Problems told *Interfax*, "There is no official or unofficial evidence that there were instances of sexual intercourse or the carrying out of sexual experiments in space. At least in the history of Russian or Soviet space exploration, this most certainly was not the case."[1]

Despite the many denials by NASA and Roscosmos, the rumors and conspiracy theories continue. For many, the denials are part of one huge government coverup. In 1997, a former NASA consultant wrote a book in which he claimed that the pool used for training astronauts for spacewalks was used several times for "extracurricular" activity. About the same time, French science writer Pierre Kohler wrote in his book *The Last Mission* that NASA had secretly commissioned a study on performing ten sexual positions in space aboard the Space Shuttle *Challenger* in 1996.[2]

In the book, Kohler claimed that he found evidence that "the results were videotaped but were so sensitive that the scientists only released censored copies to the space agency."[3]

Kohler's expose was picked up by several news outlets, which ran with the "gotcha" story. There was only one problem. Much to the embarrassment of the journalists working the story, the claims were nothing but a hoax. NASA spokesman Ed Campion responded with a terse "We are not, have not, and do not plan to conduct sex experiments."[4] Simultaneously, other news outlets were quick to use the internet to do some research, something Kohler apparently had not done, and found that the story was based on a bogus internet post that had previously been debunked.

Whether these conspiracy theorists actually believe that the government is hiding something from the public or whether they have voyeur-like tendencies and only wish that there had been sex in space so they can hear all of the titillating details, the fact of the matter is that there are many reasons why there has never been a romantic hookup in space. The bottom line is, as of right now, sex in space cannot be the sensuous, Earth-moving experience many imagine it to be. Which leads us to some interesting tales about attempts to study how men and women will navigate the tricky business of weightless (let us dispense with the cutesy word play now) "close encounters" to join the "250-Mile Club" or "Big Bang."

With longer duration missions to Mars and beyond and the potential colonization of other worlds, the subject of human sexual relationships becomes a serious discussion, not only for procreation, but also to fulfill the basic human sexual needs and desires of the men and women who will be confined to a small spacecraft for extended periods of time. As NASA's Chief Medical Officer for Spaceflight Arnauld Nicogossian simply put it, "One day, it will happen."[5]

Reproduction in space was one of the topics briefly discussed during a congressional hearing in 1962 to determine if women faced discrimination by NASA when it came to hiring them as astronauts and whether they were "suited" for spaceflight. Two of the women who had trained to be astronauts in the early 1960s, Jerrie Cobb and Jane Hart of the Mercury 13, were asked to appear before the US House of Representatives Special Subcommittee on the Selection of Astronauts. During the hearing, New York Congressman Victor Anfuso acted like a giddy schoolboy, sliding a sexual innuendo into the conversation:

> Anfuso: Miss Cobb, I think that we can safely say at this time that the whole purpose of space exploration is to someday colonize these other planets and I don't see how we can do that without women. [Laughter] . . . I call on Mrs. Hart.
>
> Mrs. Hart: I would like to say, I couldn't help but notice that you call upon me immediately after you referred to colonizing space.
>
> Mr. Anfuso: That's why I did it. [Laughter][6]

And that's how the topic of sex in space has always been treated, either with an embarrassed, blushing giggle or as a taboo subject that should not be discussed by decent people. But it is a serious subject that one day will have to be examined. Many factors must be considered, but one thing is certain: having an intimate encounter in zero gravity is not as easy as one might think. As the late American author, actress, and space advocate Vanna Bonta once said, "Sex in space is more than just a big bang."[7]

The first hurdle is the issue of Sir Isaac Newton's third law of motion: if object A exerts a force on object B, object B also exerts an equal and opposite force on object A,[8] or what you may remember from your high school days as "for every action there is an equal and opposite reaction." When talking about sexual activity in a microgravity or weightless environment, that means that a couple's gyrations would shoot the passionate lovers across the spacecraft in opposite directions from one another. And even if they did manage to hang on, vehicles like the International Space Station have fans that help circulate air that would push the lovers around like a puck on an air hockey table.

The simplest solution would be to hold each other tightly, right? Easier said than done, as Bonta and her husband discovered firsthand. Besides being an author, Vanna Bonta was deep into the science of space exploration, advocating for more daring space ventures, including commercial travel and eventual colonization of the planets. In 2007, she participated in NASA and Northrup Grumman's Lunar Lander Challenge, in which contestants were tasked with building a lightweight commercial landing vehicle for a return mission to the moon. Bonta entered her design for a pressure-release device that would be used on high-combustion engines.[9]

Bonta's fascination with space extended into researching ways for a couple to meet one of the most basic of human needs while weightless. She designed a special suit that would allow a couple to cling to each other called 2Suit. The idea came to her in 2006 after she and her husband were invited by the National Space Society to ride on a modified Boeing 747 named *G-Force One* (affectionately nicknamed the "Vomit Comet"), in which the pilot takes the plane to a specific altitude then performs several arcs or parabolas that cause the occupants inside to float weightless for up to thirty seconds at the top of each arc.

In a documentary produced by the History Channel discussing the pros and cons of fooling around in space, Bonta stressed how difficult even the simplest of romantic gestures became. "We somersaulted, we spun around, we flew, we did various acrobatics, but the one impression that I really had was just how zero, zero-gravity really is and that there is no attraction. You just couldn't—you know—you really had to work at it just for a kiss."[10]

The suit she designed was made of lightweight material, and the jacket fronts were lined with Velcro. When unzipped, they allowed the individuals to attach to the other's suit, providing close face-to-face contact. Describing the suit, Bonta stressed that it was not only for sex.

"It was a suit that you could hook up with your partner and you didn't have to work at staying close or hanging out, and it's not just for sexual activity but intimacy. The feeling of intimacy and closeness, of physical proximity—cuddling, cozy, hanging out."[11]

On December 3, 2008, two prototype 2Suits were readied as Vanna and her husband once again took to the skies in *G-Force One*. Their goal was simple enough—complete one successful "docking" (i.e., simply attaching the two suits together) and a kiss before gravity resumed. During several attempts, the couple, eyes wide open as if afraid of crashing into the other, did just that, colliding without successfully linking the suits. Finally, on the eighth try, it was mission accomplished. The husband-and-wife team's suits joined the couple together tightly in midair, where they gave one another a kiss then sailed to the floor of the plane as gravity returned. Bonta was satisfied that the test was a success.

Other than actually being able to hold one another during a rendezvous, there are several other factors—and dangers—that must be considered and overcome before space becomes home to the new Love Shack. The first is motion sickness. Many astronauts, when they first become weightless after launch, feel the effects of motion sickness as their bodies try to orient themselves in their new environment, but just as quickly, they adapt and all is well. A floating tryst, on the other hand, creates plenty of motion, which could bring on nausea, and the results would not be pretty. So much for romance.

Passionate romance also can get steamy—literally. Beads of sweat won't roll down a body, but rather float around the astronauts, pooling around their bodies.

Scientists also hold differing opinions about how blood flow affects male arousal. Studies have shown that an astronaut's blood pressure is lower in space, since it doesn't have to fight gravity. Micro- or zero gravity also causes the human body to lose muscle mass, which would make the act of lovemaking extremely tiring.

Then there is the problem of reproduction. It is believed that microgravity or no gravity at all can alter sperm count and its behavior. Radiation is one of the biggest issues. If not shielded properly in space, the human body is susceptible to a constant bombardment of subatomic particles. On Earth, we are protected from 99 percent of space radiation by the planet's atmosphere and magnetic field. In space, those particles could damage DNA, causing cancer or other dangerous side effects. Additionally, human reproductive organs are extremely sensitive to radiation and easily damaged.

Although no human has gotten lucky in space, several experiments with rats were conducted by Russian cosmonauts. In the study, some of the female rats stopped ovulating, estrogen levels in both male and female rats dropped so low that they weren't even interested in the opposite sex, and the corpus luteum, which is responsible for producing hormones that protect an embryo until the placenta can grow enough to do its job, was severely damaged in a few of the female rats.[12] Even studies on plants aboard the International Space Station have shown that conception is impacted by the microgravity environment.

As shown, intimacy among humans in space is complex with potential perils that need to be studied further. State University of New York human sexuality instructor Ray Noonan emphasizes the real need to study the issue: "Besides the stress-reduction and exercise aspects," Noonan says, "it could have morale-building features. For most people, sexual relationships tend to enhance their happiness and performance."

Aside from the joys of experiencing the ultimate climax, the possibility of pregnancy exists. *Sex in Space* author Laura Woodmansee strongly opposes any star-struck couple having sex in space until more research is done about the implications, both psychologically and biologically, especially when it comes to conception. Woodmansee believes that star-struck lovers who want to try a sexual encounter in orbit should be informed with warnings like those on a pack of cigarettes: "Don't get pregnant in space."

"The research that has come out today on plant sex and conception in space," Woodmansee wrote in an opinion piece for Space.com, "highlights the fact that we simply don't know the impact space conditions would have on human conception and pregnancy. Right now, it would be unethical to conceive a baby in orbit or even risk conception. That's my bottom line."[13]

That bottom line makes fooling around two hundred miles or more from Earth sound like a frightening nightmare instead of a stimulating, romantic experience. But it isn't stopping some creative entrepreneurs from going boldly where no man (and woman) has gone before. Enter the adult film industry and, once again, a financially strapped Russian space program.

As mentioned in chapter 12, once the checkered flag was lowered on the space race with the United States, the Russian space program found it difficult to fund further missions and projects. Roscosmos, the Russian space agency, began coming up with some highly creative methods of raising money and ensuring that its spacecraft, cosmonauts, and the Mir space station could keep flying. Mir was heavily used for advertising and marketing purposes. At one point, the station nearly became the first orbital movie studio, taking lovemaking to new heights.

In 2000, Roscosmos granted permission for a movie to be filmed aboard the space station. The movie would be an adaptation of Russian author Chingiz Aytmatov's book, *The Mark of Cassandra*.[14] The story was quite appropriate for Mir: a renegade cosmonaut is determined to stay onboard an aging Russian space station until the end of the station's life and its ultimate destruction when it reenters Earth's atmosphere. In the adaptation, a female space commander is sent to the station to persuade the crew member to return to Earth before it's too late.[15] You see where this is heading.

"This is, of course, an exotic project," Roscosmo Director Yuri Koptev said in an interview, "but the [agency] considers it possible in order to get additional money."[16] Another spokesman for the agency, Sergei Gorbunov, added, "Everything depends on technical conditions, including the actual state of Mir."[17]

The script called for the two cosmonauts to have a super sexy—and super erotic—male-female encounter aboard the station. In an article published by *Wired*, it was rumored that Emma Thompson and Willem Dafoe were approached to star in the movie,[18] although that has never been confirmed. In any event, Russian actor Vladimir Steklov eventually was selected for the lead male role and both Natalia Gromushkina and Olga Kabo were considered for the female role. In the end, Gromushkina won the part, and with casting complete, the actors underwent a series of medical tests, which they passed before basic cosmonaut training was to begin.

Before the project was able to get off the ground, as it were, the plug was pulled without warning. The film's director, Yuri Kara, could not secure funding for the project. Not long after, much like in the screenplay, Mir was allowed to die a peaceful death, burning up in Earth's atmosphere in 2001 over the South Pacific (with no one onboard), and the movie idea was forever shelved.[19]

The dream of being the first to produce an adult movie in space was not dead, however, and an unnamed triple-X film company approached the brash and adventurous billionaire and entrepreneur Sir Richard Branson to make it happen.

Branson dreamed of making space travel commercially viable, and in 2004, his fledgling company, Virgin Galactic, was born. The centerpiece of the venture was a winged aircraft, much like the space shuttle, which would fly a pilot, copilot, and six passengers to an altitude of up to fifty miles above Earth, or what is known as suborbital flight. The original plan was to have the first test flight with Branson onboard flying out of Sierra County, New Mexico, in 2009, but it wasn't until July 12, 2021, that he made his dream come true by flying just beyond the edge of space.

In 2008, Virgin Galactic was approached with an offer to use the company's spacecraft to film a sex video with a couple experiencing the ultimate afterglow. The president of Virgin Galactic, Will Whitehorn, told reporters that the offer came from an unidentified party. "[It] was for $1 million, up front, for a sex-in-space movie. That was money we had to refuse."[20]

Seven years later in 2015, the idea of a sexy space movie was revived, this time by the online pornography website, PornHub. To raise the needed $3.5 million to produce such a movie, PornHub set up a crowd funding page[21] and made its case to the public, relating the project to historic spaceflights of the past and hinting that the movie would be made in the name of "research."

> Without great explorers and adventurers, the world as we know it would be a completely different place. Be it by the discovery of new lands or even by way of industrial and cultural innovation, great minds and brave souls have forever changed the way that we see and experience the world. Columbus, Galileo, Da Vinci, Edison, and Ford, among others, have all physically and culturally helped shape the planet that we currently call home. Recently, however, the focus has been shifting over and out to what lies beyond the Earth's atmosphere. Some are contemplating colonizing Mars while others are promising elevators into space by 2050. One way or another, there are many elements about life in space that need careful consideration and research . . . especially sex. As such, PornHub is teaming up with top ranking adult studio Digital Playground in joining the ranks of Armstrong and Gagarin by pioneering a one of a kind mission to defy gravity, make history, and push the boundaries of intergalactic "Sexploration" by filming the first ever sex tape in space. In doing so, we will not only be changing the face of the adult industry, we will also be chronicling how a core component of human life operates while in orbit.[22]

Money raised from crowdfunding would go toward researching and developing the equipment needed for such a venture, researching the mechanics of shooting such a movie in a suborbital weightless environment, and sending the movie's stars to an accelerated six-month astronaut training camp.[23]

The company assured donors that it would be money well spent, confirming that the project was 100 percent real. The first step was to cast the movie. Adult film stars Eva Lovia and Johnny Sins were inked to play the "sextronauts." Incentives for those who donated to the project included an advance screening for a $10 donation, a video chat with the stars of the movie for $20, and the actual spacesuits worn by the stars for a $1,500 donation.[24]

In an interview with the Huffington Post, PornHub Vice President Corey Price detailed the proposed flight and scene to be shot. "The production of the video will begin when the spaceship takes flight. Filming will commence upon takeoff and as the ship climbs, so too will the lovemaking. . . . As soon as the ship reaches its maximum altitude, [they] will be weightless for at least a few minutes. Our actors will be having sex and climaxing in that time frame. Ideally, of course."[25]

As soon as the crowdfunding page went live on the website IndieGoGo, complete with a short film promoting the movie by the stars, who winked and dropped a few innuendoes, the company raised $1,000 from forty-four donors but, ultimately, the numbers never added up. Once the sixty-day fund-raising campaign was over, the project had netted only $236,086 from 1,388 backers, only 6 percent of the goal. Despite fanfare from the mainstream media about the project, the fundraiser quietly shut down, and with it, the last known attempt at making an adult movie in space.

As mankind moves ever deeper into space and as commercial enterprises begin to race toward the heavens to make space travel as routine as commuting to work, humans having intimate relationships in interstellar space and on other planets will happen eventually. If we are to leave this world to explore and inhabit other worlds, it is inevitable. As Vanna Bonta also said, "Sex in space is not just a good idea, it's survival."[26]

EARTH IS THE CRADLE OF HUMANITY, BUT ONE CANNOT LIVE IN A CRADLE FOREVER

It is incredible to think that it was only four years between the launch of Sputnik 1 in 1957 and the launch of the first man in space. What's even more remarkable is the fact that it was only twelve years after Sputnik that a man walked on the moon. Today, mankind has a continual presence in space, whether it is orbiting Earth aboard the International Space Station, flying unmanned spacecraft to distant planets, or even touching asteroids millions of miles away. Still, this history is only a blip in the human historical record, a speck of dust in comparison to human existence on the planet.

If there is one thing that the story of mankind's exploration of space can tell us, it is that Earth is a precious place. Following a successful suborbital flight and landing in the west Texas desert, billionaire Jeff Bezos and the crew of *New Shepard* were asked what the flight meant to them. The answer was one that the world had heard many times before from the select group of men and women who had flown in space before them. "This is the only good planet in the solar system," Bezos told reporters, "and we have to take care of it. . . . [W]hen you go into space and see how fragile it is, you'll want to take care of it even more."[1]

All of our astronauts and cosmonauts return to Earth with the same feeling: that the planet is one in a million. There is no planet B if we don't take care of it. No one expressed this feeling more succinctly than astronomer, astrophysicist, and planetary scientist Carl Sagan.

In early 1990, Sagan watched in awe as a manmade probe, Voyager 1, raced toward the edge of our solar system and interstellar space, the first such probe to do so. At the suggestion of Sagan, the Jet Propulsion Laboratory turned the spacecraft around and sent the command to take a picture. When the image returned to Earth, the scene was stark and put our precious lives into perspective. In the grainy photo, surrounded by inky blackness and shafts of rainbow colors from the sun, there was a dot. A pale blue dot. That was us.

"The Earth is a very small stage in a vast cosmic arena," Sagan wrote of the photo.

> Think of the rivers of blood spilled by all those generals and emperors so that, in glory and triumph, they could become the momentary masters of a fraction of a dot. Think of the endless cruelties visited by the inhabitants of one corner of this pixel on the scarcely distinguishable inhabitants of some other corner, how frequent their misunderstandings, how eager they are to kill one another, how fervent their hatreds.
>
> Our posturing, our imagined self-importance, the delusion that we have some privileged position in the Universe, are challenged by this point of pale light. Our planet is a lonely speck in the great enveloping cosmic dark. . . . To me, it underscores our responsibility to deal more kindly with one another, to preserve and cherish the pale blue dot, the only home we've ever known.[2]

Although many others have echoed those words over the years, the reality is that we are not good stewards of our pale blue dot. Whether you believe in climate change or not, it is undeniable that Mother Earth is angry, a mother scolding her children. Temperatures are rising causing drought, intense and uncontrollable wildfires, and suffering. Hurricanes are increasing in frequency and strength.

Although NASA is known for having its eyes looking skyward, it always has had one eye firmly trained on the planet itself, monitoring Earth's health like a doctor caring for a patient in a hospital ICU bed.

NASA's involvement with earth science began early in the agency's history with the launch of America's first weather satellite, the Television Infrared Observation Satellite (TIROS) 1 on April 1, 1960. Five hours after launch, the satellite began sending back telemetry and images, including incredible photos of a typhoon, which aided meteorologists in unlocking answers to several questions about the planet's weather patterns but added many more that needed to be answered.

Between 1960 and 1965, ten TIROS satellites were launched, each building on the knowledge gained from its predecessor. From there, NASA went on to help pioneer computer modeling that would enhance understanding and predicting the planet's climate and weather processes.

It's not only the climate that NASA keeps a close eye on. Satellites help monitor the carbon in our oceans, track movement and behavior of marine and land-based animals, track water quality, study water conservation and the health of rainforests, and more.

The space agency also practices what it preaches on the ground. In the shadow of the world's largest rocket, the space launch system that will return men and women to the moon, the 140,000-acre Merritt Island National Wildlife Refuge provides protected habitat for 330 species of birds, endangered sea turtles, manatees, and more. The refuge and the adjoining Canaveral National Seashore are managed by the US Fish and Wildlife Service and Department of Interior.

Taking conservation and Earth research a step further, NASA and the National Oceanic and Atmospheric Administration (NOAA) teamed up in 1986 to build Aquarius, the world's first and only undersea research laboratory where "aquanauts" study the ocean and its marine life and where astronauts train for space travel sixty-two feet below the surface as part of the NEEMO project.[3]

NEEMO 22's international crew "splashed down" to the undersea Aquarius laboratory on the floor of the Atlantic Ocean on June 18, 2017, for a ten-day mission. *NASA*

Whether in space or under the sea, one thing is certain. No matter what the future holds for life here on Earth or for human exploration, there will be more tales to be told of heroic missions, inspiring people, and those special side stories that take us down the back alleys of history and make mankind's adventures in space—and in life—extra special and fun. There are volumes of blank pages just waiting to be filled. The journey continues.

NOTES

CHAPTER 1. THE ROCKET WORKED PERFECTLY EXCEPT FOR LANDING ON THE WRONG PLANET

1. "Brief History of Rockets," Glenn Research Center, June 12, 2014, www.grc
.nasa.gov/www/k-12/TRC/Rockets/history_of_rockets.html.

2. L. Carrington Goodrich and Fêng Chia-shêng, "The Early Development of Firearms in China," *Isis* 36, no. 2 (1946): 114–23, www.jstor.org/stable/225875.

3. "Brief History of Rockets," Glenn Research Center.

4. Steve Carper, *Max Valier: Rocket Man* (Rochester, NY: Farstream Books, 2015), 91.

5. "Fantastic Speed Car," *Dundee Courier*, April 13, 1928, www.britishnewspa
perarchive.co.uk/viewer/bl/0000564/19280413/108/0007.

6. "Von Braun: Germany," Colorado Space Grant Consortium, September 19, 2006, https://spacegrant.colorado.edu/COSGC_Projects/space/PastClasses/01
_Spring/GEEN_2850_4850/Presentations/Class_07_History_Koehler/Rocket%
20History/vonBraun/Wernher%20Von%20Braun%20Germany.htm.

7. George D. Morgan, *Rocket Age: The Race to the Moon and What It Took to Get There* (Guilford, CT: Prometheus Books), 23.

8. *Spaceflight*, episode 2, "Thunder in the Skies," directed by Blaine Baggett, aired 1985, Public Broadcasting Service, https://youtu.be/IqK-QN7iP98.

9. "Story of Explorer 1," Marshall Spaceflight Center, January 19, 2018, www
.nasa.gov/centers/marshall/history/explorer1/explorer-1.html.

10. Ian MacDougal, "The Leak Prosecution That Lost the Space Race," *The Atlantic*, August 8, 2016, www.theatlantic.com/politics/archive/2016/08/the-leak -prosecution-that-lost-the-space-race/495659/.

11. Milton Bracker, "Vanguard Rocket Burns on Beach," *New York Times*, December 7, 1957, https://timesmachine.nytimes.com/timesmachine/1957/12/07 /87342556.html?auth=login-smartlock&pageNumber=1.

12. "Russians at U.N. Tweak US on (Satellite) Nose," *New York Times*, December 7, 1957, 8, https://timesmachine.nytimes.com/timesmachine/1957/12/07/ issue.html?auth=login-smartlock.

CHAPTER 2. HEY SKY, TAKE OFF YOUR HAT, I'M ON MY WAY

1. Elaine F. Weiss, "Before Rosie the Riveter, Farmerettes Went to Work," *Smithsonian Magazine*, May 28, 2009, www.smithsonianmag.com/history/before -rosie-the-riveter-farmerettes-went-to-work-141638628/.

2. Kelly A. Spring, "Women's Land Army of World War I," National Women's History Museum, 2017, www.womenshistory.org/resources/general/womens-land -army-world-war-i#:~:text=In%201917%2C%20the%20United%20States,which% 20included%20Britain%20and%20France.&text=During%20World%20War%20 I%2C%20Britain,of%20the%20country's%20agricultural%20sector.

3. "Rosies Kept America Running during World War II," USO Stories, December 27, 2015, www.uso.org/stories/106-rosies-kept-america-running-during-world -war-ii.

4. Richard Jennings and Stephen Vernonneau, "Remembering Dr. William Randolph Lovelace II," *Aerospace Medicine and Human Performance* 89, no. 5 (2018): 491–92, www.researchgate.net/publication/324867159_Remembering _Dr_William_Randolph_Randy_Lovelace_II/link/5b101d114585150a0a5d723b /download.

5. Robert B. Voas, "Astronaut Training," in *Mercury Project Summary (NASA SP-45)*, NASA History Archive, https://history.nasa.gov/SP-45/ch10.htm.

6. Stephanie Nolen, *Promised the Moon: The Untold Story of the First Women in the Space Race* (New York: Thunder's Mouth Press), 26.

7. Nolen, *Promised the Moon*, 27.

8. C-SPAN, "Oral Histories: Wally Funk," July 18, 1999, video, 51:57, www .c-span.org/video/?291543-1/wally-funk-oral-history-interview&playEvent.

9. Kathy L. Ryan, Jack A. Loeppky, and Donald E. Kilgore, "A Forgotten Moment in Physiology: The Lovelace Woman in Space Program (1960–1962),"

Advances in Physiology Education 33, no. 3 (2009), https://journals.physiology.org /doi/full/10.1152/advan.00034.2009.

10. C-SPAN, "Oral Histories: Wally Funk."

11. Margaret A. Weitekamp, "NASA's Early Stand on Women Astronauts: No Present Plans to Include Women on Space Flights," Smithsonian National Air and Space Museum, March 17, 2016, https://airandspace.si.edu/stories/editorial/nasas -early-stand-women-astronauts-%E2%80%9Cno-present-plans-include-women -space-flights%E2%80%9D.

12. "Betty Skelton," Smithsonian National Air and Space Museum, https:// airandspace.si.edu/explore-and-learn/topics/women-in-aviation/Skelton.cfm.

13. "Betty Skelton," Smithsonian National Air and Space Museum.

14. "Betty Skelton," Smithsonian National Air and Space Museum.

15. *Qualifications for Astronauts: Hearings before the Special Subcommittee on the Selection of Astronauts of the Committee on Science and Astronautics*, 88th Cong., 2nd Session, July 17–18, 1962, 8–9, https://babel.hathitrust.org/cgi/pt?id=ucl .a0000094904&view=1up&seq=12.

16. *Qualifications for Astronauts*, 6.

17. *Qualifications for Astronauts*, 67.

18. Uliana Malashenko, "The First Group of Female Cosmonauts Were Trained to Conquer the Final Frontier," *Smithsonian Magazine*, April 12, 2019, www.smithsonianmag.com/science-nature/first-group-female-cosmonauts-trained -conquer-final-frontier-180971900/.

19. Malashenko, "The First Group of Female Cosmonauts."

20. Malashenko, "The First Group of Female Cosmonauts."

21. Boris Chertok, *Rockets and People*, vol. 3, *Hot Days of the Cold War* (Washington, DC: US Government Printing Office, 2009), 220.

22. Mark Wade, "Vostok 6," Astronautix.com, www.astronautix.com/v/vostok6 .html.

23. Chertok, *Rockets and People*, 3:220.

24. Chertok, *Rockets and People*, 3:226.

25. NASA, "Nichelle Nichols Shuttle Astronaut Recruitment," University of North Texas Digital Library, 1977, https://digital.library.unt.edu/ark:/67531 /metadc1770594/m1/.

26. Barbara Galloway, "January 29, 1986: Judith Resnick Remembered As Brilliant, Strong-Willed," *Akron Beacon Journal*, January 27, 2020, www.beacon journal.com/news/20200127/jan-29-1986-judith-resnik-remembered-as-brilliant -strong-willed.

27. *Written Testimony of Dr. Mae C. Jemison, United States Committee Senate Committee on Health, Education, Labor, and Pensions Hearing on "Forty Years and*

Counting: The Triumphs of Title IX," 88th Cong., 2nd Session, June 19, 2012, 2, www.help.senate.gov/imo/media/doc/Jemison.pdf.

28. Rheana Murray, "Pioneering Woman Wally Funk Was Supposed to Go to Space in the 60s. At 82, She's Getting Her Chance," Today.com, www. today.com/news/wally-funk-82-join-jeff-bezos-blue-origin-spaceflight-t224369 ?icid=canonical_related.

CHAPTER 3. IN MEMORY OF LAIKA

1. Colin Burgess and Chris Dubbs, *Animals in Space: From Research Rockets to the Space Shuttle* (New York: Springer, 2007), 41.

2. Matt Williams, "How Many Dogs Have Been to Space?" Phys.Org, October 3, 2016, https://phys.org/news/2016-10-dogs-space.html#:~:text=And%20 what%20of%20%22Man's%20Best,humanity%20a%20space%2Dfaring%20race!.

3. Chris Dubbs, *Space Dogs: Pioneers of Space Travel* (Lincoln, NE: iUniverse, 2003), 33–34.

4. "JFK Jr. on Being a Kennedy," CNN.com, www.cnn.com/videos/us /2018/02/19/1995-john-f-kennedy-jr-larry-king-live-interview-sot.cnn.

5. Richard Hollingham, "The Stray Dogs That Led the Space Race," BBC .com, November 1, 2017, www.bbc.com/future/article/20171027-the-stray-dogs -that-paved-the-way-to-the-stars.

6. Alice George, "The Sad, Sad Story of Laika, the Space Dog, and Her One-Way Trip to Orbit," *Smithsonian Magazine*, April 11, 2018, www.smithso nianmag.com/smithsonian-institution/sad-story-laika-space-dog-and-her-one-way -trip-orbit-1-180968728/.

7. "Rocket Launching Tripled Heartbeat of Satellite's Dog," *New York Times*, March 25, 1958, 14, https://timesmachine.nytimes.com/timesmachine/1958/03/26 /issue.html.

8. "1957: Russians Launch Dog into Space," BBC, http://news.bbc.co.uk /onthisday/hi/dates/stories/november/3/newsid_3191000/3191083.stm.

9. "Laika, Dog of Year," *New York Times*, November 24, 1957, 54, https://times machine.nytimes.com/timesmachine/1957/11/24/96029488.html?pageNumber=54.

10. F. L. van der Wal and W. D. Young, "Project MIA (Mouse-in-Able), Experiments on Physiological Response to Space Flights," *American Rocket Society* 29, no. 10 (October 1959): 716–20, https://arc.aiaa.org/doi/10.2514/8.4879.

11. "Before Human Flight," Smithsonian Air and Space Museum, https:// airandspace.si.edu/exhibitions/apollo-to-the-moon/online/racing-to-space/before -human-flight.cfm.

12. "Corona and the Cold War: A Light in the Darkness," Smithsonian Air and Space Museum, March 7, 2013, https://airandspace.si.edu/exhibitions/space-race/online/sec400/sec440.htm.

13. "The U2 Dragon Lady," Lockheed Corporation, www.lockheedmartin.com/en-us/news/features/history/u2.html.

14. Andrew LaPage, "The First Discoverer Missions: America's Original (Secret) Satellite Program," DrewExMachina, April 13, 2019, www.drewexmachina.com/2019/04/13/the-first-discoverer-missions-americas-original-secret-satellite-program/.

15. Kevin C. Ruffner, "Corona: America's First Satellite Program," CIA *Cold War Records*, 1995, 16, www.cia.gov/library/center-for-the-study-of-intelligence/csi-publications/books-and-monographs/corona.pdf.

16. Ruffner, "Corona: America's First Satellite Program," 17.

17. *One Small Step: The Story of the Space Chimps*, DVD, directed by David Cassidy and Kristin Davy (Gainesville: University of Florida, 2002).

18. Rachel Gall, "This Month in Physics History—February 6, 1971: Alan Shepard Hits a Golf Ball on the Moon," *APS News* 26, no 2 (February 2017): 2.

19. Edward C. Burks, "Space Biologist Suing on Monkey-Test 'Libel,'" *New York Times*, January 20, 1973, https://timesmachine.nytimes.com/timesmachine/1973/01/20/79833233.html?pageNumber=33.

CHAPTER 4. FROM THE MOON, INTERNATIONAL POLITICS SEEM PETTY

1. Louis de Gouyon Matignon, "Edward Makuka Nkoloso, the Afronauts and the Zambian Space Program," Space Legal Issues, March 4, 2019, last modified June 18, 2019, www.spacelegalissues.com/space-law-edward-makuka-nkoloso-the-afronauts-and-the-zambian-space-program/.

2. Matignon, "Edward Makuka Nkoloso."

3. "Zambia's Forgotten Space Program," *Zambian Observer*, March 8, 2020, www.zambianobserver.com/zambias-forgotten-space-program/.

4. Namwali Serpell, "The Zambian 'Afronaut' Who Wanted to Join the Space Race," *The New Yorker*, March 11, 2017, last modified August 17, 2017, www.newyorker.com/culture/culture-desk/the-zambian-afronaut-who-wanted-to-join-the-space-race.

5. Serpell, "The Zambian 'Afronaut.'"

6. Matignon, "Edward Makuka Nkoloso."

7. "List of Space Agencies in Africa," Space in Africa, last modified July 16, 2020, https://africanews.space/list-of-space-agencies-in-africa/.

8. Louis de Gouyon Matignon, "Ariel 1, the First British Satellite," Space Legal Issues, April 13, 2019, last modified April 19, 2019, www.spacelegalissues.com/ariel-1-the-first-british-satellite/.

9. Papers of John F. Kennedy Presidential Papers National Security Files, "Space Activities: US/USSR cooperation 1961–1963," www.jfklibrary.org/asset-viewer/archives/JFKNSF/308/JFKNSF-308-006.

10. National Aeronautics and Space Administration NASA History Office, "The Decision to Go to the Moon: President John F. Kennedy's May 25, 1961, Speech before a Joint Session of Congress," https://history.nasa.gov/moondec.html.

11. Nikita Sergeyevich Khrushchev to John F. Kennedy, February 21, 1962, as printed in US Congress, Senate, Committee on Aeronautical and Space Sciences, *Documents on International Aspects of the Exploration and Use of Outer Space, 1954–1962*, 88th Cong., 1st sess., 1963, 232, https://history.nasa.gov/SP-4209/ch2-2.htm#source5.

12. Office of the Historian, "Letter from President Kennedy to Chairman Khrushchev," https://history.state.gov/historicaldocuments/frus1961-63v06/d41.

13. Edward Clinton Ezell and Linda Nueman Ezell, "The First Dryden-Blagonravov Agreement—1962," in *The Partnership: A History of the Apollo-Soyuz Test Project*, NASA History Archive, https://history.nasa.gov/SP-4209/ch2-3.htm.

14. President John F. Kennedy, "Address before the 18th General Assembly of the United Nations, September 20, 1963," John F. Kennedy Presidential Library and Museum, www.jfklibrary.org/archives/other-resources/john-f-kennedy-speeches/united-nations-19630920.

15. Edward Clinton Ezell and Linda Nueman Ezell, "The Dryden-Blagonravov Talks—1964–1965," in *The Partnership: A History of the Apollo-Soyuz Test Project*, NASA History Archive, https://history.nasa.gov/SP-4209/ch2-5.htm.

16. "Alouette I and II," Government of Canada, last modified September 28, 2018, www.asc-csa.gc.ca/eng/satellites/alouette.asp.

17. "Treaty on Principles Governing the Activities of States in the Exploration and Use of Outer Space, including the Moon and Other Celestial Bodies," United Nations Office for Outer Space Affairs, www.unoosa.org/oosa/en/ourwork/spacelaw/treaties/introouterspacetreaty.html.

18. NASA Space Science Data Coordinated Archive, "Luna 1—NSSDCA/COSPAR ID 19519-012A," National Aeronautics and Space Administration, https://nssdc.gsfc.nasa.gov/nmc/spacecraft/display.action?id=1959-012A.

19. Joel Achenbach, "50 Years after Apollo, Conspiracy Theorists Are Still Howling at the 'Moon Hoax,'" *Washington Post*, May 24, 2019, last updated July

16, 2019, www.washingtonpost.com/national/health-science/50-years-after-apollo
-conspiracy-theorists-are-still-howling-at-the-moon-hoax/2019/05/23/ca5b4a3a
-700e-11e9-9f06-5fc2ee80027a_story.html.

20. Olivia McKelvey, "Conspiracy Theorist Punched by Buzz Aldrin Still In-sists Moon Landing Was Fake," *Florida Today*, July 19, 2019, last updated July 23, 2019, www.floridatoday.com/story/tech/science/space/2019/07/19/lunar-landing
-denier-we-never-went-moon/1702676001/.

21. Charles Arthur, "USSR Planned to Atom Bomb the Moon," *Independent*, July 9, 1999, last updated October 22, 2011, www.independent.co.uk/news/ussr
-planned-to-atom-bomb-moon-1105344.html.

22. Percy Greg, *Across the Zodiac* (Salt Lake City, UT: Guttenberg Press, 2003), www.gutenberg.org/cache/epub/10165/pg10165.html.

23. William Harris and Nathan Chandler, "How Astronauts Work," How-StuffWorks, last modified August 25, 2020, https://science.howstuffworks.com
/astronaut1.htm.

24. "Russians Coin a Word for Him: 'Cosmonaut,'" *New York Times*, April 13, 1961, 16, https://timesmachine.nytimes.com/timesmachine/1961/04/13/10145
6133.html?pageNumber=16.

CHAPTER 5. $12 A DAY TO FEED AN ASTRONAUT: WE CAN FEED A STARVING CHILD FOR $8

1. Barry Mitchell, "Sonny Fox, a Wonderama Guy!" October 18, 2009, video, 3:42, https://youtu.be/ltywPIWDohY.

2. William Sims Bainbridge, "The Impact of Space Exploration on Public Opin-ions, Attitudes, and Beliefs," in *Historical Studies in the Societal Impact of Space-flight*, ed. Steven J. Dick (Washington DC: National Aeronautics and Space Ad-ministration Office of Communication, 2016), 12, www.nasa.gov/sites/default/files
/atoms/files/historical-studies-societal-impact-spaceflight-ebook_tagged.pdf.

3. Margot Lee Shetterly, "Katherine Johnson Biography," NASA: From Hid-den to Modern Figures, last updated February 24, 2020, www.nasa.gov/content
/katherine-johnson-biography.

4. Margot Lee Shetterly, "Mary W. Jackson Biography," NASA: From Hidden to Modern Figures, last updated February 8, 2021, www.nasa.gov/content/mary
-w-jackson-biography.

5. Richard Paul and Steven Moss, *We Could Not Fail: The First African Ameri-cans in the Space Program* (Austin: University of Texas Press, 2015), 272.

6. Paul and Moss, *We Could Not Fail*, 272.

7. Paul and Moss, *We Could Not Fail*, 354.

8. "Central High School: A Cold War Hot Spot," University of Arkansas Little Rock Center for Arkansas History and Culture, https://ualrexhibits.org/legacy/cold-war/.

9. "Central High School."

10. John Kirk, "Sputnik 1," Public Radio from University of Arkansas Little Rock, News and Culture for Arkansas, October 5, 2017, www.ualrpublicradio.org/post/sputnik-1.

11. Kirk, "Sputnik 1."

12. Mark Garcia, "60 Years Ago: The U.S. Response to Sputnik," NASA History Office, November 16, 2017, www.nasa.gov/feature/60-years-ago-the-us-response-to-sputnik/.

13. Bob Granath, "NASA Helped Kick-Start Diversity in Employment Opportunities," NASA's Kennedy Space Center, July 1, 2016, www.nasa.gov/feature/nasa-helped-kick-start-diversity-in-employment-opportunities.

14. Steven L. Ross, "NASA and Racial Equality in the South, 1961–1968" (PhD diss., Texas Tech University, 1997).

15. "2 Negroes Barred from Alabama U.," *New York Times*, March 26, 1963, https://timesmachine.nytimes.com/timesmachine/1963/03/26/86610496.html?pageNumber=1.

16. Paul and Moss, *We Could Not Fail*, 128.

17. John Lewis, *Walking with the Wind: A Memoir of the Movement* (New York: Simon & Schuster, 1998), 138.

18. "Two Challenges: On Rights and Space," *New York Times*, March 21, 1965, https://timesmachine.nytimes.com/timesmachine/1965/03/21/96701034.html?pageNumber=166.

19. Ben A. Franklin, "Wallace Is Given a NASA Warning," *New York Times*, June 9, 1965, https://timesmachine.nytimes.com/timesmachine/1965/06/09/issue.html.

20. *Hearings before the Subcommittee on Executive Reorganization of the Committee of Government Operations*, 89th Congress, Second Session, December 14 and 15, 1966, Part 14, 2–3, https://college.cengage.com/history/ayers_primary_sources/king_justice_1966.htm.

21. "NASA Chief Briefs Abernathy after Protest at Cape," *United Press International*, July 16, 1969, www.upi.com/Archives/1969/07/16/NASA-chief-briefs-Abernathy-after-protest-at-Cape/7371558396299/.

22. Thomas O. Paine, "Memorandum for Record" (Washington, DC: National Aeronautics and Space Administration, 1969).

23. Paine, "Memorandum for Record."

24. Paine, "Memorandum for Record."

25. "NASA Chief Briefs Abernathy."

26. Andrew Chaikin, "Live from the Moon: The Societal Impact of Apollo," in *Historical Studies in the Societal Impact of Spaceflight*, ed. Steven J. Dick (Washington, DC: National Aeronautics and Space Administration Office of External Affairs, 2007), 4, https://history.nasa.gov/sp4801-chapter4.pdf.

27. Marcus Baram, *Gil Scott-Heron: Pieces of a Man* (New York: St. Martin's Press, 2014), 75.

28. "Pruning the Apollo Program," NASA History Office, 1978, www.hq.nasa .gov/office/pao/History/SP-4204/ch22-8.html.

CHAPTER 6. IN MY DAY, THE BIGGEST THING YOU COULD HAVE DONE WAS BECOME A SECRETARY

1. "Three Black Astronauts Share Their Small Steps, Giant Leaps: Three Black Astronauts Share Their Story," NBC News, April 30, 2016, last updated May 13, 2016, www.nbcnews.com/news/nbcblk/three-black-female-astronauts-share -their-small-steps-giant-leaps-n546641.

2. Robert Stone and Alan Andres, *Chasing the Moon: The People, the Promise, and the Politics That Launched America Into Space* (New York: Ballentine, 2019), 102.

3. "Cleric Told NASA Selects Astronauts on Ability Only," *Baltimore Afro-American*, March 31, 1962, 3.

4. Emily Mathay and Stacey Flores Chandler, "Patient No Longer: Fighting for Representation in Space," John F. Kennedy Presidential Library and Museum, last updated March 30, 2021, https://jfk.blogs.archives.gov/2019/06/12/represen tation-in-the-space-race/.

5. "Adam Yarmolinsky: Kennedy, Johnson Advisor," *Los Angeles Times*, January 12, 2000, www.latimes.com/archives/la-xpm-2000-jan-12-mn-53307-story.html.

6. Smithsonian Channel, "Black in Space: Breaking the Color Barrier," February 1, 2020, video, 51:18, https://youtu.be/I7jJ8jEh608.

7. Richard Paul and Steven Moss, *We Could Not Fail: The First African Americans in the Space Program* (Austin: University of Texas Press, 2015), 90.

8. Steven L. Ross, "NASA and Racial Equality in the South, 1961–1968" (PhD diss., Texas Tech University, 1997), 58, https://ttu-ir.tdl.org/handle/2346/17967.

9. Smithsonian Channel, "Black in Space."

10. Smithsonian Channel, "Black in Space."

11. Shareef Jackson, "Ed Dwight Was Going to Be the First African American in Space. Until He Wasn't," *Smithsonian Magazine*, February 18, 2020,

www.smithsonianmag.com/history/ed-dwight-first-african-american-space-until
-wasnt-180974215/.

12. Smithsonian Channel, "Black in Space."

13. Charles L. Sanders, "The Troubles of 'Astronaut' Edward Dwight," *Ebony*,
March 1965, 32.

14. General Chuck Yeager and Leo Janos, *Yeager: An Autobiography* (Toronto:
Bantam Books, 1985), 306, https://archive.org/details/yeagerautobiogra00yeag
/mode/2up?q=dwight.

15. Smithsonian Channel, "Black in Space."

16. "Negro One of 15 in Space Course," *New York Times*, April 1, 1963,
48, https://timesmachine.nytimes.com/timesmachine/1963/04/01/89525941.html
?pageNumber=48.

17. Stone and Andres, *Chasing the Moon*, 123.

18. Smithsonian Channel, "Black in Space."

19. Adrian Horton, "Breaking the Color Barrier: Behind the Long Fight to
Diversify Space," *Guardian*, February 24, 2020, www.theguardian.com/tv-and-ra
dio/2020/feb/24/black-in-space-breaking-the-color-barrier-nasa-space-race-interview.

20. *New York Times*, "I Was Poised to Be the First Black Astronaut," December
19, 2019, video, 12:38, https://youtu.be/Xj1sJQW98nE.

21. Jon Uri, "50 Years Ago: NASA Benefits from Manned Orbiting Laboratory
Cancellation," NASA History Archive, June 10, 2019, www.nasa.gov/feature/50
-years-ago-nasa-benefits-from-mol-cancellation.

22. Mark Garcia, "Robert Lawrence: First African-American Astronaut," NASA
History Office, February 21, 2018, www.nasa.gov/feature/robert-lawrence-first
-african-american-astronaut.

23. Carl A. Posey, "A Sudden Loss of Altitude," *Air & Space Magazine*, July
1998, www.airspacemag.com/space/a-sudden-loss-of-altitude-14456179/.

24. Smithsonian Channel, "Black in Space."

25. T. A. Heppenheimer, "The Turn of Congress," in *The Space Shuttle Decision: NASA's Search for a Reusable Space Vehicle*, NASA History Archive, https://
history.nasa.gov/SP-4221/ch4.htm.

26. "The Intercosmos Program," Yu.A. Gagarin Research and Test Cosmonaut
Training Center, www.gctc.su/main.php?id=235.

27. NASA Content Administrator, "Guy Bluford Remembered 30 Years Later,"
NASA History Office, August 26, 2013, last updated August 7, 2017, www.nasa
.gov/vision/space/workinginspace/bluford_1st_african_amer.html.

28. NASA Content Administrator, "Guy Bluford Remembered."

CHAPTER 7. THE PROBABILITY OF SUCCESS IS DIFFICULT TO ESTIMATE

1. Bill Safire, memorandum to H. R. Haldeman, "In the Event of Moon Disaster," National Archives, July 18, 1969, www.archives.gov/files/presidential -libraries/events/centennials/nixon/images/exhibit/rn100-6-1-2.pdf.

2. Nola Taylor Redd, "What if Apollo 11 Failed? President Nixon Had Speech Ready," Space.com, May 10, 2014, www.space.com/26604-apollo-11-failure -nixon-speech.html.

3. William Safire, "Essay; Disaster Never Came," *New York Times*, June 12, 1999, 15, www.nytimes.com/1999/07/12/opinion/essay-disaster-never-came.html.

4. Leo Sartori and Kosta Tsipis, *Phillip Morrison, 1915–2005: A Biographical Memoir* (Washington, DC: National Academy of Sciences, 2009), 13.

5. Lloyd S. Swenson Jr., James M. Grimwood, and Charles C. Alexander, *This New Ocean: A History of Project Mercury* (Washington, DC: NASA Special Publication, 1989), 429, https://history.nasa.gov/SP-4201/ch13-4.htm.

6. Swenson, Grimwood, and Alexander, *This New Ocean*, 432.

7. Swenson, Grimwood, and Alexander, *This New* Ocean, 432.

8. Barton C. Hacker and James M. Grimwood, *On the Shoulders of Titans: A History of Project Gemini* (Washington, DC: NASA Special Publication, 1977), 268, https://history.nasa.gov/SP-4203/ch12-2.htm.

9. D. C. Agle, "Riding the Titan II," Space.com, September 1998, www.air spacemag.com/flight-today/riding-the-titan-ii-223545/.

10. C-SPAN, "Walter Schirra Oral History Interview," December 1, 1988, video, 44:23, www.c-span.org/video/?455845-1/walter-schirra-oral-history-interview.

11. Fran Foley, "Interview with Walter M. Schirra Jr.," Library of Congress Veterans History Project, April 19, 2004, http://memory.loc.gov/diglib/vhp/story /loc.natlib.afc2001001.12840/transcript%3FID%3Dsr0001.

12. Paul Recer, "Shuttle Narrowly Misses Ocean Ditching during Launch," Associated Press, July 30, 1985, https://apnews.com/article/f70d7c808a559ac0c425fc 04cc1dc55a.

13. William J. Broad, "Challenger Limps into a Low Orbit as an Engine Fails," *New York Times*, 1, https://timesmachine.nytimes.com/timesmachine/1985 /07/30/156467.html?pageNumber=1.

14. Recer, "Shuttle Narrowly Misses Ocean."

15. John Uri, "45 Years Ago: Apollo-Soyuz Test Project L-3 Months," NASA, April 15, 2020, www.nasa.gov/feature/45-years-ago-apollo-soyuz-test-project-l-3 -months.

16. Uri, "45 Years Ago: Apollo-Soyuz."

17. Nigel Packham, Scott Johnson, Dennis Pate, Patrick Huckaby, Joanna Opaskar, and Faisal Ali, "Significant Incidents and Close Calls in Human Space-flight," NASA Safety and Mission Assurance, September 30, 2019, https://sma .nasa.gov/SignificantIncidents/.

18. Rex D. Hall and David J. Shayler, *Soyuz, a Universal Spacecraft* (Chichester, UK: Praxis Publishing, 2003), 305, https://books.google.com/books?id=db Gchpi1HP8C&q=soyuz+t-10#v=snippet&q=soyuz%20t-10&f=false.

19. Hall and Shayler, *Soyuz, a Universal Spacecraft*, 306.

20. Hall and Shayler, *Soyuz, a Universal Spacecraft*, 306.

21. Federal Aviation Administration, *Returning from Space* (Washington, DC: Government Publishing Office, n.d.): 4, www.faa.gov/about/office_org/headquar ters_offices/avs/offices/aam/cami/library/online_libraries/aerospace_medicine/tu torial/media/iii.4.1.7_returning_from_space.pdf.

22. Charles Q. Choi, "New Spaceship Antenna Prevents Radio Silence during Fiery Reentry," Space.com, May 2015, www.space.com/29675-hypersonic-space craft-communications-reentry-tech.html.

23. *Secret Space Disasters: Gasping for Air*, directed by Paul Epstein, Anuar Arroyo, and Simon Martin, aired 2016 on the Science Channel, www.discoveryplus .com/video/secret-space-escapes/gasping-for-air.

24. "Stafford Says All 3 Astronauts at Fault in Apollo Gas Leak," *New York Times*, August 10, 1975, 32, www.nytimes.com/1975/08/10/archives/stafford -says-all-3-astronauts-are-at-fault-in-apollo-gas-leak.html.

25. *Secret Space Disasters*.

26. *Secret Space Disasters*.

27. Arnauld E. Nicogossian, *The Apollo-Soyuz Test Project Medical Report* (Washington, DC: Scientific and Technical Information Office National Aeronautics and Space Administration, 1977), 32.

28. "Stafford Says All 3 Astronauts at Fault."

CHAPTER 8. GOOD MORNING TO OUR BEAUTIFUL WORLD AND TO ALL THE BEAUTIFUL PEOPLE WHO CALL IT HOME

1. Abigail Harrison, "What Time Is It on the International Space Station," AstronautAbby.com, May 3, 2013, www.astronautabby.com/the-international -space-station-time/.

2. Colin Fries, *Chronology of Wakeup Calls*, NASA History Office, March 3, 2015, https://history.nasa.gov/wakeup%20calls.pdf.

3. "Music to Wake Up By," NASA History Office, August 10, 2005, www.nasa.gov/vision/space/features/wakeup_calls.html.

4. Matt Blitz, "How the NASA Wake-Up Call Went from an Inside Joke to a Beloved Tradition," April 27, 2017, www.popularmechanics.com/space/a26229/nasa-wake-up-call/.

5. Howard Taubman, "Space Men Heard Earthling Songs," *New York Times*, February 7, 1968, 43, https://timesmachine.nytimes.com/timesmachine/1968/02/07/90027006.html?pageNumber=43.

6. Fries, *Chronology of Wakeup Calls*.

7. Fries, *Chronology of Wakeup Calls*.

8. Fries, *Chronology of Wakeup Calls*.

9. "William Shatner Gives Space Shuttle Discovery Wake-Up Call before Return to Earth," *New York Post*, March 7, 2011, https://nypost.com/2011/03/07/william-shatner-gives-space-shuttle-discovery-wake-up-call-before-return-to-earth-video/.

10. NASA Johnson, "STS-26 Wakeup Call: Robin Williams and Space Shuttle Discovery," August 12, 2014, video, 1:48, www.youtube.com/watch?v=S7WJtQYU8i4.

11. MuppetWiki, "Space Shuttle Columbia: Pigs in Space," February 24, 2016, video, 2:55, https://youtu.be/Fa900RnL_aI.

12. VideoFromSpace, "SpaceX Demo-2 Astronaut's Kids Wake Up Dads in Adorable Splashdown Day Call," video, 1:31, https://youtu.be/tJxY1CyZdhw.

13. Marcia Dunn, "Astronauts: SpaceX Dragon Capsule 'Came Alive' on Descent," WNYF-TV, August 4, 2020, www.wwnytv.com/2020/08/04/astronauts-spacex-dragon-capsule-came-alive-descent/.

14. "Actor Paul Reiser," Geffen Playhouse, www.geffenplayhouse.org/people/paul-reiser/.

15. Fries, *Chronology of Wakeup Calls*.

16. Taubman, "Space Men Heard Earthling Songs."

17. "Bells, Gemini 6," Smithsonian Air and Space Museum, https://airandspace.si.edu/collection-objects/bells-gemini-6/nasm_A19670148001.

18. "This Day in History: Gemini VI Crew Pull Christmas Prank, First Song Performed in Space," Space Coast Daily, December 16, 2017, https://spacecoastdaily.com/2017/12/this-day-in-history-gemini-vi-crew-pull-christmas-prank-first-song-performed-in-space/.

19. "Jingle Bells Tinkle As Astronauts Play a Duet on Gemini 6," *New York Times*, December 17, 1965, 28, https://timesmachine.nytimes.com/timesmachine/1965/12/17/95010530.html?pageNumber=28.

20. CBC Music, "Chris Hadfield and Barenaked Ladies | I.S.S. (Is Somebody Singing)," February 12, 2013, video, 5:28, https://youtu.be/AvAnfi8WpVE.

21. Channel 5 News, "Commander Chris Hadfield: David Bowie Really Liked How I Portrayed Space Oddity," December 13, 2013, video, 6:06, https://youtu .be/_rTcIpWxy9I.

22. Becky Ferreira, "Chris Hadfield's Spirited Song in Space Was No Oddity," *New York Times*, November 2, 2020, www.nytimes.com/2020/11/02/science /chris-hadfield-space-oddity.html.

23. Stephanie Schierholz, "NASA Astronaut Cady Coleman, Jethro Tull's Ian Anderson Perform First Space-Earth Flute Duet," NASA, April 11, 2011, www .nasa.gov/home/hqnews/2011/apr/HQ_11-108_Coleman_space_duet.html.

24. MIT Alumni Channel, "NASA Astronaut Cady Coleman '83 on Losing Flute on the ISS," December 12, 2014, video, 1:21, https://youtu.be/B77FNP3nyZA.

25. Jethro Tull and Ian Anderson, "Ian Anderson + Cady Coleman Flute Duet in Space," April 8, 2011, video, 2:19, https://youtu.be/XeC4nqBB5BM.

26. AP Archive, "Update: Sir Paul McCartney Serenades Space Crew," July 21, 2015, video, 6:27, https://youtu.be/j7pXWv_VLlk.

27. Lloyd S. Swenson, James M. Grimwood, Charles C. Alexander, *This New Ocean—The Official History of Project Mercury* (St. Petersburg, FL: Random House, 2010), 352, https://history.nasa.gov/SP-4201/ch11-3.htm.

28. Wilson Rothman, "Pre-Launch Jitters and Then . . . Liftoff," Gizmodo.com, May 6, 2009, https://gizmodo.com/pre-launch-jitters-and-then-liftoff-5241957.

CHAPTER 9. HI, IT'S DOUG AND BOB AND WE'RE IN THE OCEAN

1. Gary Jordan, "Welcome Home, Bob and Doug," Houston We Have a Podcast, August 28, 2020, www.nasa.gov/johnson/HWHAP/welcome-home-bob-and -doug.

2. Robert Z. Pearlman, "50 Years Ago, Wally Schirra Piloted 'Sigma 7' into Space," October 3, 2012, www.nbcnews.com/id/wbna49274995.

3. Fran Foley, "Interview with Walter M. Schirra Jr.," April 19, 2004, http://memory.loc.gov/diglib/vhp/story/loc.natlib.afc2001001.12840/transcript% 3FID%3Dsr0001.

4. Bedford Brass Quintet, "Gemini 6: Jingle Bells," March 15, 2013, www .youtube.com/watch?v=RmsOmqf7Hso.

5. Owen Edwards, "The Day Two Astronauts Said They Saw a U.F.O. Wearing a Red Suit," *Smithsonian Magazine*, December 2002, https://www.smithsonian mag.com/history/day-two-astronauts-said-they-saw-ufo-santa-suit-109444898/.

6. "Tom Stafford's Jingle Bells and Wally Schirra's Harmonica," Smithsonian National Air and Space Museum, December 16, 2014, https://airandspace.si.edu /stories/editorial/tom-staffords-jingle-bells-and-wally-schirras-harmonica.

7. "Walter Schirra," Johnson Space Center Oral History Project, produced by Johnson Space Center History Office (Houston: C-SPAN American History TV, 1998), video, 17:42, www.c-span.org/video/?455845-1/walter-schirra-oral-history -interview.

8. Ulli Lotzmann, "Playmate No. 2 and EVA-2 Surveyor Activities (Continued)," 2001, JPG, 120kB, www.hq.nasa.gov/alsj/a12/a12.lmpcuf08.jpg.

9. D. C. Agle, "Playmates on the Moon," *Playboy Magazine*, December 1994, 138, www.hq.nasa.gov/alsj/a12/cuff12.html.

10. Agle, "Playmates on the Moon."

11. Eric Berger, "How a Thanksgiving Day Gag Ruffled Feathers in Mission Control," ARS Technica, November 23, 2020, https://arstechnica.com/science/2020/11 /that-time-on-thanksgiving-when-debris-threatened-the-space-shuttle/.

12. Berger, "How a Thanksgiving Day Gag."

13. Berger, "How a Thanksgiving Day Gag."

14. Berger, "How a Thanksgiving Day Gag."

15. Berger, "How a Thanksgiving Day Gag."

16. Rick Houston and Milt Heflin, *Go, Flight! The Unsung Heroes of Mission Control, 1965–1992* (Lincoln: University of Nebraska Press, 2015), 13.

17. Kevin M. Rusnak, "Owen K. Garriott Interviewed by Kevin M. Rusnak Houston, Texas—6 November 2000," NASA Johnson Space Center Oral History Project, November 6, 2000, https://historycollection.jsc.nasa.gov/JSCHistoryPor tal/history/oral_histories/GarriottOK/GarriottOK_11-6-00.htm.

18. "Tell Me a Story: Owen Garriott's 'Gotcha' during Skylab 3," Kennedy Space Center Visitor Complex, April 17, 2019, https://youtu.be/6XRuEvxWjEk.

19. "The Ancient and Honorable Order of the Turtle," Ancient Turtle Order, https://ancientturtleorder.webs.com/turtle-guide.

20. Wally Schirra, "Gotchas and Turtles," WallySchirra.com, www.wallysch irra.com/gotcha.htm.

21. Schirra, "Gotchas and Turtles."

22. "Apollo 10 Onboard Voice Transcription," Manned Spacecraft Center, June 1969, 34, www.hq.nasa.gov/alsj/a410/AS10_CM.PDF.

23. "Apollo 10 Onboard Voice Transcription," 28.

24. "Apollo 10 Onboard Voice Transcription," 5.

25. Amy Shira Teitel, "How NASA Kept Astronauts from Swearing on the Moon," Gizmodo.com, January 1, 2012, https://io9.gizmodo.com/how-nasa-kept -astronauts-from-swearing-on-the-moon-5873762.

26. Roger Simmons, "Foul-Mouthed Apollo Astronauts Got Space Program in Trouble 50 Years Ago," June 17, 2019, www.military.com/off-duty/2019/06/17 /foul-mouthed-apollo-astronauts-got-space-program-trouble-50-years-ago.html.

27. Apollo 16 Debriefing, 1969, NASA Headquarters, MP3, 3:40, www.hq .nasa.gov/alsj/a16/a16a1284635.mp3.

28. Apollo 16 Debriefing, 6:15.

29. Teri Maddox, "She Had a Deed for Land on the Moon in 1969, but NASA Wouldn't Give Her Any Rocks," *Belleville News-Democrat*, July 23, 1969, www .bnd.com/news/local/article232977347.html.

30. Maddox, "She Had a Deed for Land on the Moon."

31. Maddox, "She Had a Deed for Land on the Moon."

32. Maddox, "She Had a Deed for Land on the Moon."

33. Jess Zimmerman, "People Have Been Claiming to Own the Moon for over 250 Years," Atlas Obscura, November 12, 2015, www.atlasobscura.com/articles /people-have-been-claiming-to-own-the-moon-for-over-250-years.

CHAPTER 10. "JUST IN CASE" IS THE CURSE OF PACKING

1. "Code of Federal Regulations (Annual Edition)," US Government Publication Office, January 1, 2020, www.govinfo.gov/app/collection/cfr/2020/.

2. National Aeronautics and Space Administration, "Aeronautics and Space," *Code of Federal Regulations*, title 1214 (2004 comp): 601(c), www.govinfo.gov /content/pkg/CFR-2004-title14-vol5/pdf/CFR-2004-title14-vol5-sec1214-603.pdf.

3. NASA, "Aeronautics and Space," 601(a).

4. NASA, "Aeronautics and Space," 604(c).

5. NASA, "Aeronautics and Space," 604(c).

6. Wally Schirra, "Gotchas and Turtles," WallySchirra.com, www.wallysch irra.com/gotcha.htm.

7. Catherine Harwood, "Frank Borman Interviewed by Catherine Harwood, Las Cruces, New Mexico—13 April 1999," NASA Johnson Space Center Oral History Project, April 13, 1999, https://historycollection.jsc.nasa.gov/JSCHistory Portal/history/oral_histories/BormanF/Bormanff_4-13-99.htm.

8. Chris Carberry, *Alcohol in Space* (Jefferson, NC: McFarland, 2019), 1127.

9. Madalyn Murray O'Hair et al. v. Thomas O. Paine et al., 312 F. Supp. 434 (1969), (Civ A. No. A-69-CA-109), https://law.justia.com/cases/federal/district -courts/FSupp/312/434/1468840/.

10. Encyclopedia Britannica, "Apollo 11: Celebrating Communion on the Moon," July 8, 2019, video, 5:26, www.youtube.com/watch?v=UQRvceotTMo.

11. Carberry, *Alcohol in Space*, 1155.

12. "Webster Presbyterian Church History," Webster Presbyterian Church, www.websterpresby.org/content.cfm?id=329.

13. John W. Young with James R. Hansen, *Forever Young: A Life of Adventure in Air and Space* (Gainesville: University of Florida Press, 2013), 83.

14. Sandra Blakeslee, "Real Sandwiches Please Spacemen," *New York Times*, May 21, 1969, 20, https://timesmachine.nytimes.com/timesmachine/1969/05/21 /90106729.html?pageNumber=20.

15. "Phase 3: Second Orbit," Space Log Gemini 3, https://gemini3.spacelog .org/page/00:01:49:03/.

16. Young and Hansen, *Forever Young*, 84.

17. Young and Hansen, *Forever Young*, 85.

18. "Mission Report: Apollo 10," Johnson Space Center, June 17, 1969, https://er.jsc.nasa.gov/seh/Ap10.html.

19. Andy Saunders, "The Mystery behind Alan Shepard's 'Moon Shot' Re- vealed," USGA Golf Museum, February 5, 2021, www.usga.org/content/usga/home -page/articles/2021/02/shepard-moon-club-50th-anniversary-usga-museum.html.

20. Al Van Helden, "On Motion," The Galileo Project, 1995, http://galileo.rice .edu/sci/theories/on_motion.html.

21. Exploring Stamps, "The Apollo 15 Scandal—S3E6," March 22, 2019, video, 15:31, https://youtu.be/yrkY_HB4ZS8.

22. Harold M. Schmeck Jr., "Apollo 15 Crew Reprimanded," *New York Times*, July 12, 1972, https://timesmachine.nytimes.com/timesmachine/1972/07/12/8079 6364.html?pageNumber=1.

23. "Moon Mail," *Smithsonian*, July 7, 2020, https://postalmuseum.si.edu /exhibition/stamps-take-flight-rarities-and-special-holdings/moon-mail.

24. Schmeck, "Apollo 15 Crew Reprimanded."

25. Schmeck, "Apollo 15 Crew Reprimanded."

26. Julie Mianecki, "Apollo 15's Al Worden on Space and Scandal," *Smithson- ian Magazine*, October 2011, www.smithsonianmag.com/arts-culture/apollo-15s -al-worden-on-space-and-scandal-73346679/.

27. Exploring Stamps, "The Apollo 15 Scandal."

28. Abigail Rosenthal, "Here Are Some of the Personal Items the Apollo 11 Astronauts Took to the Moon," *Palm Beach Post*, July 16, 2019, www.palmbeach

post.com/zz/lifestyle/20190716/here-are-some-of-personal-items-apollo-11-astro
nauts-took-to-moon.

29. Glenn Garner, "NASA to Make History with Mars Helicopter—and Part
of the Wright Brothers' Plane Is Aboard," People.com, March 24, 2021, https://
people.com/human-interest/nasa-mars-helicopter-history-wright-brothers/.

30. Jessica Orwig, "Apollo 16 Astronaut Explains Hidden Message behind
the Family Portrait He Left on the Moon," *Independent*, November 2, 2015, www
.independent.co.uk/news/science/apollo-16-astronaut-explains-hidden-message
-behind-family-portrait-he-left-moon-a6718111.html.

31. Orwig, "Apollo 16 Astronaut Explains Hidden Message."

CHAPTER 11. WRECKED BY THE MOST EXPENSIVE HYPHEN IN HISTORY

1. Yi-Jin Yu, "NASA's Lucky Charm for a Successful Mission? Peanuts,"
Today.com, February 23, 2021, www.today.com/food/nasa-s-perseverance-mars
-rover-landing-aided-lucky-peanuts-t209825.

2. Kevin Wilcox, "Perseverance Team Overcomes Seven Minutes of Terror," NASA Appel Knowledge Services, February 23, 2021, https://appel.nasa
.gov/2021/02/23/perseverance-team-overcomes-seven-minutes-of-terror/.

3. "Abel 1 (Pioneer 0)," NASA Science Solar System Exploration, August 12,
2019, https://solarsystem.nasa.gov/missions/pioneer-0/in-depth/.

4. "What Are NASA's Lucky Peanuts?" NASA Science Solar System Exploration, February 19, 2021, https://solarsystem.nasa.gov/news/10022/what-are-nasas
-lucky-peanuts/.

5. "Ranger 7," NASA Science Solar System Exploration, September 10, 2019,
https://solarsystem.nasa.gov/missions/ranger-7/in-depth/.

6. "PR 29-1997: New Cassini-Huygens Launch Date," European Space
Agency, July 16, 1997, https://sci.esa.int/web/cassini-huygens/-/37236-pr-29
-1997-new-cassini-huygens-launch-date#:~:text=The%20reason%20for%20
the%20delay,Cassini%2DHuygens%20onto%20the%20launcher.

7. "Mariner," Mission and Spacecraft Library Program JPL, https://space.jpl
.nasa.gov/msl/Programs/mariner.html.

8. "Mariner 1," NASA Space Science Data Coordinated Archive, https://nssdc
.gsfc.nasa.gov/nmc/spacecraft/display.action?id=MARIN1.

9. Zachary Crockett, "The Hyphen That Destroyed a NASA Rocket," Priceo-
nomics, June 20, 2014, https://priceonomics.com/the-typo-that-destroyed-a-space
-shuttle/.

10. Gladwin Hill, "For Want of Hyphen Venus Rocket Is Lost," *New York Times*, July 28, 1962, https://timesmachine.nytimes.com/timesmachine/1962/07/28/87314569.html?auth=login-email&pageNumber=1.

11. Arthur C. Clarke, *The Promise of Space* (New York: Harper and Row, 1968).

12. "Historical Log," NASA Science Mars Exploration Program, https://mars.nasa.gov/mars-exploration/missions/historical-log/.

13. "Beagle 2," NASA Science Solar System Exploration, May 10, 2018, https://solarsystem.nasa.gov/missions/beagle-2/in-depth/.

14. "Giovanni Virginio Schiaparelli," Encyclopedia Britannica, www.britannica.com/biography/Giovanni-Virginio-Schiaparelli.

15. Leo Sartori and Kosta Tsipis, "Phillip Morrison 1915–2005: A Biographical Memoir," (Washington DC: National Academy of Sciences, 2019), 13, www.nasonline.org/publications/biographical-memoirs/memoir-pdfs/morrison-philip.pdf.

CHAPTER 12. SPACE IS OPEN FOR BUSINESS

1. Fisher Space Pen Co., "Fisher Space Pen on QVC in Space!!!," January 18, 2018, video, 11:09, www.youtube.com/watch?v=LGp9B7tSjgk.

2. "50 Years Fisher Space Pen Final," SkyDive TV, August 8, 2016, https://vimeo.com/177146505.

3. "50 Years Fisher Space Pen Final."

4. "50 Years Fisher Space Pen Final."

5. "History of Space Pens," History of Pencils, www.historyofpencils.com/writing-instruments-history/history-of-space-pens/.

6. Steve Garber, "The Fisher Space Pen," NASA History Office, August 3, 2004, https://history.nasa.gov/spacepen.html.

7. "Our Story," Fisher Space Pen, www.spacepen.com/about-us.aspx.

8. "Our Story."

9. "The History of Omega Watches," Precision Watches/Jewelry, April 11, 2018, https://precisionwatches.com/the-history-of-omega-watches/.

10. "The History of Omega Watches."

11. Robin Swithinbank, "50 Years on, the Omega Watch That Went to the Moon," *New York Times*, June 29, 2019, www.nytimes.com/2019/06/29/fashion/watches-omega-speedmaster-moonwatch-anniversary.html.

12. Alessandro Mazzardo, "The Omega Speedmaster History," Time and Watches, January 4, 2021, www.timeandwatches.com/p/history-of-omega-speedmaster.html.

13. Swithinbank, "50 Years On."

14. Mazzardo, "The Omega Speedmaster History."

15. Mark Blitz, "How NASA Made Tang Cool," *Food & Wine*, May 18, 2017, www.foodandwine.com/lifestyle/how-nasa-made-tang-cool.

16. Daily Motion, "Maj. John Glenn Jr.—Name That Tune," August 24, 2015, www.dailymotion.com/video/x32t3zq.

17. "Buzz Aldrin—I'm Going to Just Say It . . . TANG SUCKS!!!" TMZ, June 11, 2013, www.tmz.com/2013/06/11/buzz-aldrin-tang-sucks-apollo-11-ufos/.

18. Mark Memmott, "Now He Tells Us: 'Tang Sucks,' Says Apollo 11's Buzz Aldrin," NPR, June 13, 2013, www.npr.org/sections/thetwo-way/2013/06/13/191271824/now-he-tells-us-tang-sucks-says-apollo-11s-buzz-aldrin.

19. "Various: Home Shopping Network QVC with Russian Space Station Mir for Space Suit Sale," Reuters Screen Ocean, 2015, https://reuters.screenocean.com/record/988064.

20. Chris Olert, "A Pitch from Space Cosmonauts Pen Ad; On Earth, It's Space Suits," *Spokesman-Review*, February 8, 1998, www.spokesman.com/stories/1998/feb/08/a-pitch-from-space-cosmonauts-pen-ad-on-earth-its/.

21. "Cosmonauts on Mir Become QVC Pitchmen," *Los Angeles Times*, February 8, 1998, www.latimes.com/archives/la-xpm-1998-feb-08-mn-16921-story.html.

22. "First Commercial Filmed in Space," Guinness World Records, www.guinnessworldrecords.com/world-records/first-commercial-filmed-in-space.

23. Danischneor, "Milk in Space—Tnuva," February 23, 2011, video, 1:35, www.youtube.com/watch?v=cYc4BPS9wLo.

24. Richard Gibson, "Pizza Hut Chooses to Embrace a Pie-in-the-Sky Ad Strategy," *Wall Street Journal*, September 30, 1999, www.wsj.com/articles/SB938647339433252633#:~:text=Pizza%20Hut%20is%20expected%20to,living%20quarters%20in%20mid%2DNovember.

25. "Space Cola Wars at 35: When Coca Cola, Pepsi Tested Soda in Space," Space.com, August 11, 2020, www.collectspace.com/news/news-081120a-space-cola-wars-35-years.html.

26. "Space Cola Wars at 35."

27. Anika Gupta, "The Cola Wars. Smear Campaign in Space?" *Smithsonian Magazine*, August 1, 2008, www.smithsonianmag.com/smithsonian-institution/the-cola-wars-smear-campaigns-in-space-27520139/.

28. "Space Cola Wars at 35."

29. "Space Cola Wars at 35."

30. NASA FTCSC, "Carbonated Beverages in Space," NASA History Office, February 26, 2004, www.nasa.gov/audience/foreducators/5-8/features/F_Carbonated_Beverages_Space.html.

31. Janet Staihar, "The Space Sip—It Was Coke First," Associated Press, August 9, 1985, https://apnews.com/article/7393cd79843d04a075b3456883a5bc49.

32. John Christian, "Pepsi Plans to Project a Giant Ad in the Night Sky Using Cubesats," Futurism.com, April 13, 2019, https://futurism.com/pepsi-orbital-billboard-night-sky.

33. Jeff Foust, "Pepsi Drops Plans to Use Orbital Billboard," Space.com, April 16, 2019, www.space.com/pepsi-drops-orbital-billboard-plans.html.

34. P. H. Willekens and W. A. Peeters, "Space Marketing: A New Programme in Technical Education," *ESA Bulletin 94* (May 1998), www.esa.int/esapub/bulletin/bullet94/WILLEKENS.pdf.

CHAPTER 13. SEX IN SPACE IS MORE THAN JUST A BIG BANG

1. Mike Wall, "No Sex in Space, Yet, Officials Say," Space.com, April 23, 2011, www.space.com/11473-astronauts-sex-space-rumors.html.

2. Michael Cabbage, "Lust in Space: Study Tells All," *Chicago Tribune*, March 11, 2001, www.chicagotribune.com/sns-spacesex-story.html.

3. "Sex—The Final Frontier," BBC News, February 24, 2000, http://news.bbc.co.uk/2/hi/americas/655151.stm.

4. "Sex—The Final Frontier."

5. "Sex—The Final Frontier."

6. *Qualifications for Astronauts: Hearings before the Special Subcommittee on the Selection of Astronauts of the Committee on Science and Astronautics*, 88th Cong., 2nd Session, July 17–18, 1962, 8–9, https://babel.hathitrust.org/cgi/pt?id=ucl.a0000094904&view=1up&seq=12.

7. "Vanna Bonta Talks Sex in Space," Female.com, www.female.com.au/vanna-bonta-talks-sex-in-space.htm.

8. "Newton's Laws of Motion," Glenn Research Center, November 25, 2020, www1.grc.nasa.gov/beginners-guide-to-aeronautics/newtons-laws-of-motion/#:~:text=Newton's%20Third%20Law%3A%20Action%20%26%20Reaction&text=His%20third%20law%20states%20that,words%2C%20forces%20result%20from%20interactions.

9. "Vanna Bonta," Alchetron, July 2, 2021, https://alchetron.com/Vanna-Bonta.

10. Erik Thompson, *Sex in Space*, directed by Louis Tarantino (Sherman Oaks, CA: Flight 33 Productions, 2008), video, https://play.history.com/shows/the-universe/season-3/episode-4.

11. Thompson, *Sex in Space*.

12. Maggie Koerth, "Space Sex Is Serious Business," FiveThirtyEight.com, March 14, 2017, https://fivethirtyeight.com/features/space-sex-is-serious-business.

13. Laura Woodmansee, "Sex in Space: Is It Unethical to Conceive a Child Out There?" Space.com, March 13, 2013, www.space.com/20220-sex-in-space.html.

14. "Chingiz Aytmatov," Britannica.com, June 6, 2021, www.britannica.com /biography/Chingiz-Aytmatov.

15. Brian Harvey, *The Rebirth of the Russian Space Program—50 Years after Sputnik, New Frontiers* (Chichester, UK: Praxis, 2007), 29-30.

16. Cabbage, "Lust in Space."

17. "Russian Film Director Shoots for the Stars," *Wired*, December 22, 1997, www.wired.com/1997/12/russian-film-director-shoots-for-the-stars/.

18. "Russian Film Director."

19. Cabbage, "Lust in Space."

20. Peter B. de Selding, "Virgin Galactic Rejects $1 Million Space Porn," NBC News, October 2, 2008, www.nbcnews.com/id/wbna26991760.

21. The PornHub Team, "Pornhub Space Program—Sexploration," IndieGoGo .com, www.indiegogo.com/projects/pornhub-space-program-sexploration#/.

22. The PornHub Team, "PornHub Space Program."

23. Justin Wm. Moyer, "PornHub Crowdfunds for Sex Tap Filmed in Space," *Washington Post*, June 11, 2015, www.washingtonpost.com/news/morning-mix /wp/2015/06/11/pornhub-crowdfunds-for-sex-tape-filmed-in-space/.

24. Jack Guzman, "PornHub Launches Crowdfund to Film Movie in Space," CNBC, June 10, 2015, www.cnbc.com/2015/06/10/pornhub-launches-crowd fund-to-film-movie-in-space.html.

25. David Moye, "PornHub Crowdfunds First Porn Shot in Space," HuffPost, December 6, 2017, www.huffpost.com/entry/first-porn-in-space_n_7553126.

26. Alan Boyle, "Outer Space Sex Carries Complications," NBC News, July 23, 2006, www.nbcnews.com/id/wbna14002908.

CHAPTER 14. EARTH IS THE CRADLE OF HUMANITY, BUT ONE CANNOT LIVE IN A CRADLE FOREVER

1. Aylin Woodward, "Jeff Bezos Went to Space to Realize How Fragile Earth Is. A 10-minute Flight May Not Be Long Enough to Experience this 'Overview Effect,'" *Business Insider*, July 20, 2021, www.businessinsider.com/bezos-blue -origin-space-fragile-earth-overview-effect-astronauts-emotions-2021-7.

2. Carl Sagan, "A Pale Blue Dot," The Planetary Society, 1994, www.planetary .org/worlds/pale-blue-dot.

3. "NEEMO," NASA Mission Pages, www.nasa.gov/mission_pages/NEEMO /index.html.

INDEX

ABOUT THE AUTHOR

Joe Cuhaj grew up in New Jersey as a space fanatic. He would skip school to watch every launch and recovery from the late Mercury missions to the final Skylab mission, all while building and flying model rockets. Cuhaj is a navy veteran and former radio broadcaster turned author and freelance writer. He began his radio career just outside of New York City but moved to Mobile, Alabama, in 1981 with his wife, who is from the Port City. During this time, Joe worked in various positions including news director/reporter, where he applied to take part in NASA's Journalist in Space program, but he never heard back.

He is the author of nine books, including *Hiking Waterfalls Alabama*, *Baseball in Mobile*, and *Hidden History of Mobile*.